A PRACTICAL (
TO AIR QUALITY CO

A PRACTICAL GUIDE TO AIR QUALITY COMPLIANCE

SECOND EDITION

RUSSELL E. ERBES, CCM

John Wiley & Sons, Inc.
New York ■ Chichester ■ Brisbane ■ Toronto ■ Singapore

This text is printed on acid-free paper.

Library of Congress Cataloging in Publication Data:

Erbes, Russell E.
A practical guide to air quality compliance / Russell E. Erbes. — 2nd ed.
p. cm.
Includes bibliographical references and index.
ISBN 0-471-15006-1 (cloth : alk. paper)
1. Air quality management—United States. 2. Air quality—Standards—United States. 3. United States. Clean Air Act. 4. United States. Clean Air Act Amendments of 1990. I. Title.
TD883.2.E73 1996
363.73'9256'0973—dc20 95-46166

Printed in the United States of America

10 9 8 7 6 5 4 3 2 1

CONTENTS

ACKNOWLEDGMENTS

It is not often that one has the privilege of writing a book that represents his life career, but that is what I have done here. I enjoy air quality work and have been doing it for most of my professional life. When my son asks what I do at work, my reply is to say that I am trying to keep the air invisible. I believe and hope that this book will help us all achieve the goal of clean air, but achieve it in a manner consistent with economic viability.

It is also not often that one has the opportunity to publicly thank his or her spouse for wisdom, patience, support, and encouragement. If you ever have the chance to meet my wonderful wife, Mina, you will know why I say, "Thank you, Mina, for everything." This book is dedicated to her and my father, Edward Erbes.

Finally, I would like to thank all of my air quality and environmental colleagues whose own wisdom and patience have allowed me to develop and prepare this book. Among those are Charles Moore, professor emeritus, who gave me my first scientific air quality job and whose emphasis on practicality, functionality, and understanding the fundamental underlying principles provided the foundation for this book; Lou Grant, professor emeritus, who persuaded me to continue in the field of atmospheric science; and Ralph Williams, president and CEO of Class One Technical Services, who worked with me back in 1975 in my first industrial air quality job and whose review of the manuscript is greatly appreciated. My thanks also go to Isabelle Cohen, my editor at John Wiley & Sons, who has also been my cheerleader.

Thanks also go to all of you who have purchased the book and are trying to comply with the multitude of air quality requirements. I hope the book is helpful to you and that it makes your life easier. Good luck.

CAVEAT

This publication is designed to provide accurate information; however, legislation and regulations are undergoing constant change. In addition, local and state regulatory authorities have specific local interpretations that may contradict those contained herein. The publication is sold with the understanding that neither the publisher nor author is engaged in rendering legal, consulting, or other professional service. If such advice or other expert assistance is needed, the services of competent professional legal counsel and/or consultants should be sought.

BIOGRAPHY

Russell E. Erbes, CCM, is Senior Vice President and Director, Air Quality Services for Kleinfelder, Inc., a major environmental and engineering consulting firm. Mr. Erbes has over 20 years of air quality compliance experience, ranging from field technician work (dust sampling) to preparation of testimony before the U.S. Congress on the 1990 Clean Air Act. He holds a Bachelor's degree in Physics from the New Mexico Institute of Mining and Technology and a Master's degree in Atmospheric Science from Colorado State University. He is also a Certified Consulting Meteorologist of the American Meteorological Society. Mr. Erbes has a unique combination of legislative, regulatory, and technical experience gained at a wide variety of industrial facilities, and his air quality practice focuses on assisting industrial facilities in maintaining air quality compliance.

A PRACTICAL GUIDE TO AIR QUALITY COMPLIANCE

CHAPTER 1

INTRODUCTION

The U.S. federal Clean Air Act was most recently amended on November 15, 1990, for the first time in 13 years. These amendments promulgated the most comprehensive air quality legislation in the history of the United States and significantly changed the method and type of air quality regulation. Although the Clean Air Act was originally promulgated over 30 years ago and has been amended several times since, the 1990 Amendments are by far the most significant. Some would argue that the 1990 Clean Air Act Amendments will affect routine environmental management and compliance at industrial facilities more than any other single piece of environmental legislation.

The 1990 Amendments contained legislative provisions affecting an extremely broad range of air quality issues and required the U.S. Environmental Protection Agency (EPA) to issue more than 120 new regulations in the five years between 1990 and 1995. These new programs range from requirements for detailed, several hundred-page permits to operate equipment, to labeling requirements for consumer products, to a duty to investigate the potential effects of the legislation on employment. This broad range of issues and the new requirements and concepts contained in the 1990 Amendments have caused the complexity and costs of complying with the Clean Air Act to exponentially increase, estimated by the EPA at approximately $23 billion annually when the Act is fully implemented in the year 2005.

PURPOSE

In response to this complexity, the *Practical Guide to Air Quality Compliance* has been written. The *Guide* is an aid to understanding how to comply with the federal Clean Air Act (CAA) and its amendments, including the Clean Air Act Amendments (CAAA) of 1990. This *Guide* provides practical insights into how to effectively and efficiently comply with the myriad of regulatory Clean Air Act requirements.

This *Guide* is unique because it combines detailed information regarding CAA legislation and regulation with technical and practical compliance insights. For example, the *Guide* discusses the new 1990 CAA program for air toxics. But it also discusses the regulatory implementation schedule and source-specific requirements of the program as well as the new legislative program of residual risk assessment. The *Guide* also presents methods for conducting risk assessments and provides detailed chemical-specific data that allow the reader to evaluate the potential effect of the new legislation on his or her specific facility. This combination of technical, regulatory, and legislative information is what makes the *Guide* useful.

Although focused on the federal requirements, the federal CAA is implemented at the state and local levels, and local requirements may be more stringent than those contained in the federal CAA. (For simplicity, throughout this *Guide*, the term "CAA" or "the Act" will be used to refer to the entire federal Clean Air Act as amended. In addition, for convenience, a list of acronyms is included in Appendix A.) The CAA prohibits the local requirements from being less stringent than those contained in the CAA, and local requirements are often literally or functionally equivalent to the federal requirements. Therefore, a thorough understanding of the federal requirements will also yield an in-depth understanding of the local requirements.

Although this *Guide* will focus on federal requirements, common local and state requirements that differ will often be cited. In addition, the examples used throughout the *Guide* will discuss the local versus federal distinctions, if any. The primary goal of this *Guide* is to provide a fundamental understanding of the underlying concepts embodied in the CAA and how those concepts are implemented at the state and local levels. A secondary goal is to impart to the reader knowledge of some of the detailed regulatory and technical aspects of CAA implementation, which will allow the reader to gain insight into the "tricks of the trade" in the CAA arena.

The *Guide* is written primarily from the perspective of an environmental manager of an industrial facility as his or her practice relates to air quality compliance. Therefore, the insights contained in the *Guide* are practical techniques that can be routinely used on a daily basis. It focuses on issues related to stationary sources of pollution and not mobile sources, except as those mobile sources relate to stationary source regulatory programs (e.g., substitution of mobile emission reductions for stationary source increases under the nonattainment program, Chapter 9 of the *Guide*).

The *Guide* is intended to be useful to all levels of environmental professionals, from neophyte to experienced air quality professionals.

STRUCTURE

The *Guide* is written with a focus toward providing plain language descriptions and discussions of the various legislative and regulatory requirements. These requirements will be exemplified by citing actual or hypothetical examples of how the requirements affect normal industrial practices. Actual legislative and regulatory language will be cited and referenced when necessary to demonstrate the practical impact of certain provisions, but the reader is referred to other documents (e.g., CCH, 1990) for complete legislative language and to the Code of Federal Regulations (CFR Title 40) and the Federal Register (FR) for complete regulatory language.

Chapter 2 of the *Guide* provides a historical background and overview of the CAA and presents some of the new major concepts embodied in the 1990 CAA. Chapter 3 discusses how the federal requirements of the CAA are implemented at the state and local levels, and it also discusses the interrelationships of legislation, regulation, and permits. These two chapters provide a foundation for the detailed technical and regulatory discussions contained in the following chapters.

The CAA exists to provide protection for the nation's air quality, as measured by the national ambient air quality standards (NAAQS). What these standards are, how they are established, and how they are evaluated (atmospheric dispersion modeling) is presented in Chapter 4 of the *Guide*. Chapter 5 discusses another aspect of protecting the air quality resource: risk assessment and management. Risk assessment and management are used in situations when it is generally considered that there is no "safe level" of a pollutant—for example, a suspected carcinogenic chemical.

Chapters 6 and 7 discuss how the Act establishes emission limits for sources of air pollution in order to help ensure that the NAAQS can be met. Chapter 8 presents the concepts of New Source Review (NSR), also aimed at protection of the NAAQS, while Chapter 9 discusses the Act's nonattainment requirements for those areas of the United States that do not currently meet the NAAQS. Finally, Chapter 10 discusses the permitting program (including the new Title V National Operating Permit program) contained in the Act. The permitting program is designed to ensure that each device and facility that emits air pollution complies with the emission limits discussed in Chapters 6 through 9.

Chapters 11 and 12 discuss two major new programs addressed by the 1990 CAA: acid rain and stratospheric ozone protection. These two programs are singled out for discussion because they contain, respectively, a new nationwide emissions trading concept and a global protection element of the CAA that was not included prior to the 1990 Amendments.

The 1990 CAA also contained new and very significant penalty provisions, including criminal and civil penalties for noncompliance. The enforcement aspects of the CAA are discussed in Chapter 13. Chapter 14 discusses the newly strengthened transportation and general conformity program.

Chapters 15 and 16 summarize the *Guide* by presenting the overall implementation schedule for the new 1990 CAA programs and by discussing some practical suggestions for enhancing compliance. References for the entire *Guide* are provided and the Appendices contain various lists and other detailed reference information that are of use to the environmental manager. For example, Appendix A is a list of acronyms and Appendix B is a glossary of terms. Appendix C contains a description, published by EPA, of the major new 1990 CAA requirements for specific source types.

The *Guide* is intended to be as up-to-date as possible, but the CAA programs and requirements are continuously being updated and modified as the CAA is implemented. When the new requirements and/or changes significantly change the contents of the *Guide*, revised editions will be published. Finally, readers are encouraged to contact the author at their convenience if they wish to further discuss any of the concepts contained herein. The reader is also especially encouraged to contact the author to clarify or correct any errors or inconsistencies that may appear.

CHAPTER 2

HISTORY OF AIR QUALITY LEGISLATION

AIR QUALITY LEGISLATION PRIOR TO 1977

The first federal air quality legislation, the Air Pollution Control Act, was enacted in 1955. This act authorized various federal agencies, at the request of a state or local government, to investigate the causes and effects of air pollution. This Act was primarily related to research, but established a fundamental principle for clean air legislation: The states and local governments are responsible for maintaining and improving air quality within their jurisdiction.

The current CAA began with Public Law No. 88-206, enacted December 17, 1963. This act was considered to be the first major modern environmental law enacted by the U.S. Congress (CCH, 1990). Entitled the Clean Air Act of 1963, it mostly expanded the research and technical program initiated in 1955 and provided funds to the state and local governments to develop and improve their air pollution control programs. The concept of the federal government providing funds to local governments remains fundamental in today's air quality legislative and regulatory arena. (Although, as will be seen in Chapter 10 of this *Guide*, the 1990 CAA specifically requires the states and local governments to collect fees from the regulated community to further support their air quality programs.)

The 1963 Act also began the concept of federal involvement in the air quality of a region even without a request from the local government. However, the federal government's abilities to take action were limited to specific, narrow circumstances related to interstate effects. The federal government was also tasked by this Act with the development of criteria for clean air based upon potential health effects. But, at the time, the criteria were considered to be advisory to the states and not minimum standards that must be met.

The 1963 Act was amended, or additional air quality legislation was enacted, in 1967, 1970, 1973, 1975, 1977, and 1990. The 1967 Act established four major concepts that remain part of today's air quality program:

1. Establishment of air quality control regions (AQCRs)
2. Development of specific mandated air quality criteria and control technologies
3. Adoption of National Ambient Air Quality Standards (NAAQS) to protect the public health and welfare (discussed in Chapter 4)
4. Requirements for State Implementation Plans (SIPs) to ensure that the air quality in each state meets the NAAQS

The concepts of AQCR and SIPs are discussed in Chapter 3, while the other concepts are discussed in more detail in later chapters of this *Guide.*

The 1970, 1973, and 1975 Acts established national emission limits for specific equipment (where prior to 1970 only ambient air quality standards were established). The national emission limits for specific devices and pollutants were designed to help ensure that the NAAQS would be met and were established for new or modified stationary sources (termed New Source Performance Standards, or NSPS), hazardous air pollutants (termed National Emission Standards for Hazardous Air Pollutants, or NESHAPs), and new motor vehicles. (These concepts are discussed in Chapters 6 and 7.) These acts also established the role of the federal government in air quality compliance by allowing the EPA to issue administrative orders and injunctions to force compliance and provided the legislative basis for citizens to sue the government to force action.

Note that these Acts established the concept of "technology forcing" requirements. This concept requires that a specific device meet stringent emission limits that are either not yet feasible (but soon will be) or have only recently been proven feasible, regardless of the actual ambient air quality impact of the device. Technology forcing is based on the theory that facilities should control or eliminate pollution as a matter of principle, regardless of whether there is an adverse health or welfare effect associated with the facility or device.

CLEAN AIR ACT AMENDMENTS OF 1977

The 1977 Amendments (Public Law 95-95), while continuing to refine, modify, and strengthen previously enacted requirements, added two new major regulatory programs: the Nonattainment Program and the Prevention of Significant Deterioration (PSD) program. Nonattainment programs (Chapter 9) were designed to improve air quality in regions (AQCRs) that were not able to meet the NAAQS (i.e., did not attain the standards). The PSD program, discussed in detail in Chapter 8, was designed to preserve and protect air quality in regions that were already better than the NAAQS by preventing significant deterioration of the air quality in those regions. The PSD program did this by promulgating ambient air quality increments that could not be exceeded and which were much smaller than the NAAQS. Different levels of increments were established for different areas of the United States, the most stringent increments being promulgated for national parks and major wilderness areas (see Chapter 8).

1990 CAA AMENDMENTS: GENERAL PROVISIONS

The 1990 Amendments (Public Law 101-549) to the Clean Air Act were enacted on November 15, 1990 and constitute the most significant air quality legislation in the history of the United States. The enforcement provisions of the Act were severely increased, allowing for criminal and civil penalties far greater than previously contained in the air quality Acts (see Chapter 13). In addition, the Act contains not only goals and legislative programs, but detailed technical requirements as well (e.g., the Act explicitly lists chemicals deemed to be hazardous). In summary, the 1990 Amendments brought clean air legislation to a level as stringent and complicated as legislation affecting hazardous wastes in the environment [i.e., the Resource Conservation and Recovery Act (RCRA)].

The 1990 CAA phase-in period is approximately 15 years and affects small as well as large businesses and emitting devices. The two primary driving forces for the Act were concerns over acid rain and the extensive number of nonattainment areas that had not reached the NAAQS by the deadlines established in the 1977 Act. However, while addressing those two primary concerns, numerous additional issues were also addressed by the Congress.

Among the new programs are the following:

1. Legislative programs to prevent accidental as well as routine releases (only routine releases were regulated in previous Acts) of pollutants
2. A new National Operating Permit Program (discussed in Chapter 10) that moved air quality permitting from the era of obtaining a two-page permit similar to an elevator permit to the era of obtaining an air quality permit that may be more than a hundred pages long
3. A new emphasis on control technology and emission reduction for air toxics
4. Increased efforts to attain air quality standards in polluted urban areas
5. Programs to limit regional and global air pollution issues such as acid deposition and stratospheric ozone depletion
6. More severe criminal sanctions for violating the Act
7. A nationwide system of emission trading and encouraging other market-based approaches rather than traditional command and control
8. Promotion of emission reductions through energy conservation

The general findings, as written in Section 101(a) of the 1990 CAA, can be summarized as follows:

1. Much of the nation's population is located in rapidly expanding metropolitan and urban areas that generally cross the boundary lines of local jurisdictions and often extend into two or more states.
2. The growth brought about by urbanization and industrialization, including motor vehicles, has resulted in mounting dangers to the public health and welfare from air emissions.
3. Air pollution prevention and air pollution control at its source are the primary responsibilities of the states and local governments.
4. Federal financial assistance and leadership are essential for programs to prevent air pollution.

The purposes of the CAA, as described in Section 101(b) of the Act, are to:

1. Protect and enhance the quality of the nation's air resources to promote public health and welfare and the productive capacity of the population
2. Initiate and accelerate national research to achieve prevention and control of air pollution

3. Provide technical and financial assistance to the state and local governments
4. Encourage regional air pollution prevention and control programs
5. Encourage or otherwise promote reasonable governmental actions for pollution prevention

The concept of technology forcing emission reductions and control to eliminate pollution remains a primary focus for the 1990 CAA. However, new programs involving multiple governmental jurisdictions on a regional basis have been added to the Act. These multijurisdictional programs will be discussed in Chapters 8, 9, 10, 11, and 12.

1990 CAA AMENDMENTS: SPECIFIC TITLES

The 1990 Amendments are stated in seven major titles (1 through 7) and four additional smaller titles (8 through 11). The titles and their general provisions are summarized as follows:

Title I: Air Pollution Prevention and Control. Specifies requirements for state and local governments that must be taken in order to implement the entire 1990 CAA.

Title II: National Emission Standards. Establishes emission limits for mobile sources including motor vehicles, nonroad engines and vehicles, and aircraft. The title also legislates a clean fuel vehicle (e.g., electric vehicles) program.

Title III: Hazardous Air Pollutants. Lists 189 chemicals as federally mandated hazardous air pollutants, provides for control of emissions of those pollutants, and requires the EPA to promulgate regulations reducing the potential for accidental releases as well as routine releases of toxic air pollutants.

Title IV: Acid Deposition Control. Contains provisions establishing a nationwide cap on sulfur dioxide emissions from utility boilers and a national emissions trading program (discussed in Chapter 11) in order to address concerns regarding acid deposition. The title also requires a significant reduction in nitrogen oxide emissions from utility boilers.

Title V: Permits. This title legislates the national operating permit program (discussed in Chapter 10).

Title VI: Stratospheric Ozone Protection. Deals with global air quality concerns by requiring various programs to reduce and eliminate certain substances considered to be responsible for stratospheric ozone depletion.

Titles VII through XI. Contains a new program to provide visibility protection, which could have severe consequences for potential new sources; also contains a number of administrative, procedural, and miscellaneous provisions, including provisions to preserve employment and to evaluate the economic impact of the Act. Also requires a number of new research efforts to be conducted by the EPA and other agencies.

1990 CAA AMENDMENTS: GOALS

The major pollution prevention and elimination goals of the 1990 Amendments have been summarized by the EPA (ABA, 1993) as follows:

1. Bring urban air quality into attainment with the NAAQS by the deadlines set by the Act
2. Reduce emissions of air toxics by at least 75 percent (from 1987 levels)
3. Cut annual emissions of sulfur dioxide by 10 million tons (approximately 50 percent reduction) and limit future emissions to that reduced cap
4. Phase out production of ozone depleting substances (such as chlorofluorocarbons)
5. Reduce total pollutant emissions by 24 billion tons per year from 1987 levels. These reductions include 10 billion tons of sulfur dioxide reductions, 3.6 billion tons of carbon monoxide, 3.8 billion tons of nitrogen oxides emissions, 5.5 billion tons of volatile organic compound (VOC) emissions, and 0.9 billion tons of air toxics.

The 1990 CAA required EPA to issue more than 120 new regulations by 1995, an average of 24 new rules per year. Historically, the EPA issues 5 to 8 major rules a year. Obviously, the changes in the air quality regulatory program are staggering.

FUNDAMENTAL 1990 CAA CONCEPTS

Although details of the various 1990 CAA programs will be described in the following chapters of this *Guide*, it is useful to discuss some of the fundamental new initiatives contained in the 1990 CAA.

Market-Based Incentives

One of the major initiatives, which is included in several different programs, is market-based incentives for emission reductions. The theory is that if industry is given the regulatory flexibility to choose the most cost-effective method for emission reductions, greater reductions will occur at less cost than would occur under the traditional "command and control" system.

Command and control is the system whereby the local regulatory agency, in consultation and direction with the EPA, decides what types of controls should be installed on a specific facility. Once decided, the agency then writes regulations that require such controls. On the other hand, the market-based incentive system assumes that industry will automatically choose the best and most effective method. In this system, all that the agency must do in order to improve air quality is to mandate the overall goal that industry must achieve (e.g., reduce emissions 5 percent per year for 10 years) and then step back to let industry find a way to meet the goals.

The market-based incentive program is most obvious in the acid deposition program of Title IV. In this program, Congress established in the CAA a sulfur dioxide emission cap for the United States of about 10 million tons per year. Once that cap, termed a "green field cap," was established, any new source of sulfur dioxide emissions is required to obtain emission allowances from some existing source of emissions. The allowances are created by an existing source reducing emissions through installation of additional controls, fuel switching, modifying operations, or implementing any other enforceable technique. The amount of reduction can be converted into an allowance and transferred to a new source. In this manner, growth of new sources is not prevented, but the total emissions of sulfur dioxide in the United States will not increase even as growth occurs.

Emission trading between different companies is also allowed in the following:

1. The CFC phaseout rules
2. Reformulated and oxygenated gasoline rules
3. Heavy-duty engine standards and clean-fueled vehicles rules
4. Emissions substitutions created by, for example, removing older automobiles and substituting cleaner vehicles in order to obtain emission credits

The market-based incentives are also contained in air toxics regulation when a facility can obtain a six-year deferral from a command and control requirement to install Maximum Achievable Control Technology (MACT) if the facility voluntarily reduces (prior to a new rule being proposed) air toxics emissions by 90 percent for gases and 95 percent for particulates. This allows a facility the flexibility to expend resources now to control emissions, make product substitutions, or implement other techniques *prior* to the agency requirements. If this is done, the facility can obtain "credit" for the reduction from the Agency.

Facility Bubbles

In the air toxics program, facility "bubbling" is allowed. A facility bubble allows a facility to shift emissions from one device to another within the overall facility. This allows a facility to determine where it is most cost-effective and efficient to install controls or make other emission reductions. Facility bubbling is also one of the centerpieces of the national operating permit program, where a facility can change operations within the overall requirements of the permit without having to submit a permit modification.

Compliance Certification

On the other hand, in order to ensure that a facility meets its permit conditions, for the first time in air quality legislative history, a responsible corporate official was required under the 1990 CAA programs to annually certify compliance with emission limits and other permit conditions. In addition, the recordkeeping and reporting requirements and penalties for violating the provisions were made more stringent.

Intergovernmental Requirements

The 1990 CAA contains major intergovernmental programs. One program, stated in Section 176A of the Act, requires the EPA administrator to establish Interstate Transport Commissions for areas of the country where it is deemed by either the administrator or the governor of an affected state that transport of pollution from one state to another contributes significantly to violation of a NAAQS.

These commissions may make recommendations to the EPA for additional control programs to reduce air pollution. In addition, the commission may request the Administrator to issue a finding that the State Implementation Plan (see Chapter 3) is substantially inadequate to achieve attainment. The Administrator then must approve, partially approve, or disapprove the request within 18 months and must allow for public participation in making its decision. The practical impact of the EPA determining that a SIP is inadequate is that the states must rewrite the SIP, making the local regulations more stringent. If this is not accomplished, the local governments will come under sanctions, such as loss of highway and public works funds (see Chapters 3 and 4).

Similar intergovernmental groups are the Visibility Transport Commissions, established by Section 169B(c) of the Act. These commissions are established upon request to the administrator by an affected state governor or by the administrator's own action when interstate transport of pollution contributes significantly to visibility impairment in Class I areas (e.g., national parks—see Chapter 8).

The Act, in Section 169B(f), explicitly required the administrator to establish one such commission, the Grand Canyon Visibility Transport Commission. This Commission was established as required by the Act, and its final report was due to Congress by November 15, 1995. The report must recommend what measures, if any, should be taken by the administrator to remedy and prevent visibility impairment in the Grand Canyon.

The Visibility Transport Commission report could have very far-ranging implications. As required by Section 169B(d) of the Act, the report shall consider establishing clean air corridors in which additional restrictions may be required and, most importantly, where nonattainment provisions [including emission offsets for new sources and imposition of Lowest Achievable Emission Rates (LAER); see Chapter 9] could be applied. Therefore, the Grand Canyon Commission has the potential to implement the most stringent Clean Air Act provisions (i.e., nonattainment programs) over an extremely broad area of the country and in an area where the air quality is generally already very good (with respect to NAAQS). Once the final report is issued, the administrator, by law, must act within 18 months of receipt of the report to either implement, modify, or reject the recommendations.

Regulatory Negotiation

Although not required by the 1990 CAA, the EPA has adopted a rulemaking process designed to minimize the amount of litigation over proposed rules. The 1990 CAA required EPA to issue more than 55 rules in the first two years after promulgation. Although the EPA has not met that goal, numerous significant rulemakings were accomplished. In a regulatory negotiation process (often termed "reg-neg"), the EPA establishes an advisory committee of 12 to 25 members representing industry, environmental groups, the EPA technical and regulatory staff, state and local regulatory officials, and the general public (ABA, 1993). The advisory committee meets to develop consensus on a proposed rule that is then published in the Federal Register for review and comment, just as in traditional rulemaking. The difference is that the members of the advisory committee agree that they will support and not litigate the proposed rule upon which consensus has been reached. The EPA has successfully used the reg-neg process in the following:

1. Rules regulating leaks of VOCs from equipment
2. Rules regarding reformulated gasoline
3. Rules limiting air toxics emissions from coke ovens
4. Development of the core acid rain rules
5. A negotiated agreement to reduce the emissions of sulfur dioxide by 90 percent from the Navajo Generating Station, located near the Grand Canyon of Arizona

EPA is currently actively involved in reg-neg committees for a host of control technology guidance (CTG) documents related to establishing MACT standards for air toxics (see Chapter 7) and the various CFC phaseout rules (Chapter 12).

CHAPTER

3

LEGISLATIVE AND REGULATORY CONCEPTS

DIFFERENCES BETWEEN LEGISLATION AND REGULATION

The CAA is a legislative document. It establishes the laws on which the clean air regulatory program is based. In order to implement the program, the EPA is charged with developing and promulgating federal regulations that require the states and local agencies to develop and promulgate state and local regulations. In order to promulgate regulations, the states may need to enact enabling legislation at the state level that provides the legal authority for state and local governments to promulgate binding regulations. Therefore, there is often a confusing and overlapping set of federal and state legislative language and federal, state, and local regulatory language that affects a facility.

In order to understand how the various sets of language interact, one needs to keep in mind the main purpose of each set of language. Figure 3-1 shows how each set of language from federal legislation to local regulations becomes more specific and more restrictive. Theoretically, legislation is supposed to establish the goals of a program, such as acid deposition control, and regulations are supposed to establish the specifics of how the goals will be met. However, because of perceived or actual inaction on the part of the EPA and local agencies, federal legislation has become more and more specific and reads more and more like a regulation. For example, the 1990 CAA promulgated a list of 189 chemicals that are, by definition, toxic air pollutants. This list was promulgated in frustration over the fact that prior to 1990, the EPA had listed only eight chemicals as toxic in a period of 20 years (since the 1970 CAA).

Although the public is involved in development of legislation at the federal level, through the elected Congress and lobbyists, the normal environmental manager will be most involved in development of regulations. Proposed and final regulations at the federal level are initially published in the Federal Register (FR). Each year, all of the final rules are then published in the

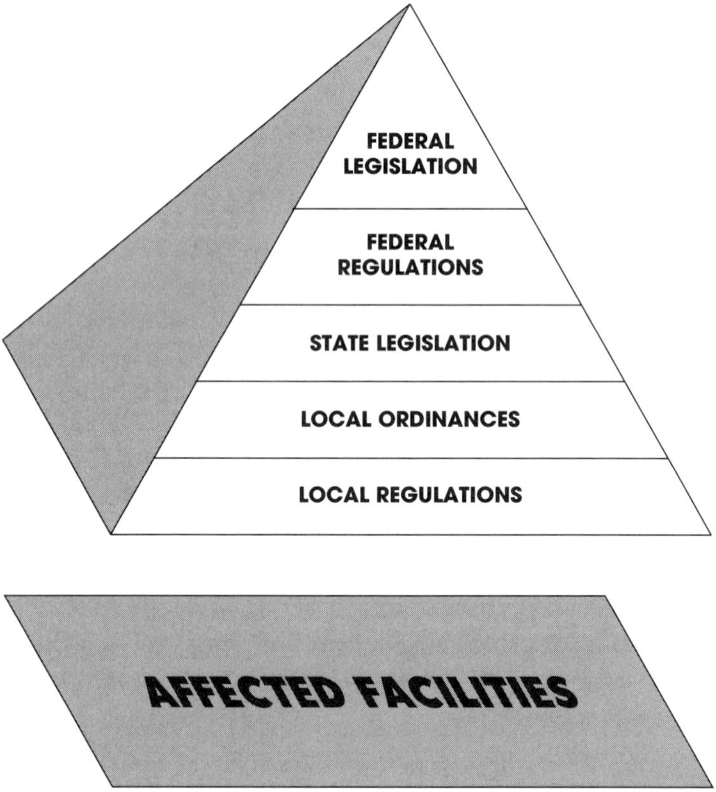

■ **FIGURE 3-1** The legislative and regulatory pyramid.

Code of Federal Regulations (CFR). Environmental regulations are in Title 40 of the CFR. The CFR is available from the U.S. Government Printing Office either by mail or at one of the several U.S. Government Printing Office stores located in major cities. The problem with the CFR is that it is revised only once a year. Although the CFR is dated January 1 of each year, it is not actually available until about nine months into the year. Therefore, it is important that the various FR promulgations of final rules be saved and referred to as well when using the CFR. Included with an FR subscription, one can obtain a document, published on a monthly basis, entitled List of Sections Affected (LSA). The LSA cross-references CFR sections to new FR final rules. The LSA greatly facilitates review of the CFR to ensure that the most current version is used.

An environmental manager should review the FR on a daily basis. The easiest way to do this is to scan the table of contents for each day's FR and make copies of only those sections of the FR that apply to the manager's specific facility. However, after selecting the appropriate portions of the FR to review, it is critical to review not only the proposed regulations, but also the background discussion related to the regulations. This background discussion is called the Preamble and describes not only what EPA is proposing, but also why and why not they are proposing specific provisions. Later, when the rule becomes final and one is attempting to comply with the rule, reading the Preamble provides great insight into what the rule is trying to accomplish. In addition, the Preamble provides references to various EPA and other publications that provide in-depth technical information related to the proposed rule. The Preamble is the easiest place to find those references. It is important to save the Federal Register proposed and final rules along with the Preamble, because once a rule is promulgated as final in the Code of Federal Regulations, only the rule is published, not the Preamble.

There are several different commercial services available that attempt to keep the environmental manager up-to-date with federal regulations. However, except for CD-ROM or on-line computer access services, the commercial services do not include all of the Preambles for both proposed and final regulations. There are services that put all of the FR and CFR on a CD-ROM disk for computer access. However, those systems are relatively expensive (about $2,000 a year while an FR subscription costs only about $350). Keeping hard copies of the relevant FR sections is an easy way to ensure that all of the information that may be needed in the future is available, although CD-ROM computer access is gaining popularity as computer search and retrieval software becomes more functional.

DIFFERENCES BETWEEN STATE AND FEDERAL LEGISLATION AND REGULATIONS

States are required by the federal Clean Air Act to implement the Act at the state and local level. States must have the legislative authority to implement the federal Act, and therefore they often have state Clean Air Acts. These Acts must give the state and local agencies sufficient authority to implement and enforce the federal Act at the state and local level. The state agencies' authority must be at least as stringent as the federal Act requires, but can, and often is, considerably more stringent.

The increase in stringency at the state level is often focused on a few specific areas in response to local political concerns. On the other hand, the state legislation can also require the state agency to enact regulations that are "no more and no less stringent" than federal regulations. Again, these "no more/no less stringent" requirements are established for specific areas where there is political concern that the state agency could become too stringent and harm the local economy.

When reviewing state legislation that contains the no more/no less stringent clauses, one must carefully determine precisely what regulations the clauses apply to. For example, New Mexico has a no more/no less stringent clause in its Clean Air Act, but the clause applies only to visibility and Prevention of Significant Deterioration (PSD) requirements (see Chapter 8) and not to other portions of the federal program. For existing sources, which are not covered by the PSD program, the state of New Mexico could and has adopted requirements that are significantly more stringent than those contained in the federal Act and EPA regulations.

A second example illustrates how the state Clean Air Act can be more stringent than the federal:

In California, the state enabling legislation is termed the California Clean Air Act. This California Act is much more stringent than the federal CAA in many areas, one of which is in how total emissions in an AQCR are treated. The California Act requires that, in nonattainment areas, the local permitting agencies (termed Air Districts in California) must establish regulatory programs that will result in an overall reduction in emissions of 5 percent per year. Many of the Air Districts are meeting this requirement by requiring *all* new sources, regardless of the magnitude of emissions, to obtain emission offsets before the new source can be constructed. (This is termed a *zero threshold offset program.*)

On the other hand, federal legislation only requires emission offsets in nonattainment areas for new sources that emit more than from 10 to 100 tons per year, depending upon the pollutant (see Chapter 9). Therefore, a new facility that plans to locate in California will face a much more stringent permitting hurdle (through the zero threshold offset requirement) than anywhere else in the nation.

To further complicate matters, however, even in California, not all of the Air Districts have implemented the California Clean Air Act in the same manner, and there are several Districts where the zero threshold offset requirement does not apply. Therefore, it is critical that an environmental manager obtain and review the local and state regulations as well as the federal regulations before making conclusions regarding air quality compliance for facilities.

STATE IMPLEMENTATION PLANS

Uses and Status

To ensure that the federal CAA is implemented at the state and local level, the federal CAA requires each state to develop and submit for approval to EPA a State Implementation Plan (SIP). The SIPs are required to detail a plan that specifies the manner in which the primary and secondary national ambient air quality standards (NAAQS) will be achieved and maintained in each AQCR of the state.

The required components of a SIP are detailed in Section 110 of the Act. The SIP is the entire collection of approved plans and regulations prepared by the state and local agencies to implement the federal requirements. Even if a local agency within a state implements the air quality program in its area, it must still submit its regulations through the state agency and those local regulations and plans become part of the SIP. Therefore, at a local level, the regulations must be approved by both the state and the EPA in order for them to become part of the SIP.

Because EPA must approve a regulation before it becomes part of the SIP, a question arises as to what happens when a regulation has been properly promulgated at the state or local level but has not yet been approved by the EPA. Simply stated, from a practical standpoint, the local regulation must be complied with regardless of the status of EPA approval. Generally, when the EPA reviews a regulation in order to approve it as part of the SIP, the EPA approves most of the regulation and disapproves only a portion of it. Once that portion is disapproved, the local agency will have to rewrite the regulation, repromulgate it, and resubmit it for approval. However, during the interim, on a local basis, the facility must still comply with the regulation as initially promulgated.

As is clear from the above process, precisely what constitutes a specific state's SIP is constantly in flux. The Federal Register almost daily contains approval decisions regarding state SIPs. Each state's formally approved SIP is described in 40 CFR 52.50 through 52.2920. Normally, exactly what is or is not included in a SIP does not affect daily operations. However, in the case of penalties and enforcement, the status of SIP approval may be important. For example, if a source is complying with a new regulation promulgated by the state (but not yet approved by the EPA as part of the SIP), but is not complying with an old regulation that *was* part of the SIP, the EPA can take enforcement action (Blattner, 1992). Therefore, if there is a compliance issue, or if a new regulation is less stringent than the old one, careful consideration of SIP approval status is important. In the new Title V permit program (Chapter 10), the SIP status is also very important.

Components of an Approvable SIP

In order for EPA to approve and include a set of regulations and plans submitted by the state in the SIP, the submittal must meet a number of conditions established by Section 110 of the federal Act and EPA regulations promulgated in 40 CFR Part 51, Appendix V. These components and implications are discussed in the following paragraphs.

Reasonable Notice and Public Hearing Before a SIP submittal is approvable, the state must provide reasonable opportunity for public notice and hearings. This includes opportunity for local political subdivisions (e.g., cities) to participate in development and discuss the proposed SIP changes prior to submittal to the EPA. Participation in this notice and hearing process by all members of the public is very important. Most state and local agencies hold public workshops and discussion sessions while developing a regulation and plan prior to the hearing. As discussed later in this chapter, effective participation in these meetings and hearings will greatly benefit a facility.

Enforceable Limitations and Controls The SIP must be enforceable by the state and include emission limitations and control requirements, including economic incentives such as marketable permits (see Chapter 11) and auctions of emission rights (often called *emission reduction credits;* see Chapter 9). The SIP must also contain provisions for obtaining data on emissions and ambient air quality and allow for those data to be provided to EPA.

This concept of enforceable limitations is fundamental to all air quality regulations. The local agencies are often sympathetic to the problems of an industry, but if a technique to solve the problem cannot be found and easily written into a regulation, the local agency will not be able to submit the regulation to the EPA.

New and Modified Source Review The SIP must contain provisions to ensure that construction of a new source or modification of an existing source will not interfere with maintenance of NAAQS. This includes prohibiting sources from contributing significantly to nonattainment in the region under jurisdiction of the local agency as well as nearby states.

Sufficient Personnel and Funding The SIP submittal must demonstrate that the state and local agencies have sufficient personnel and resources to effectively carry out implementation of their SIP. The SIP submittal must also demonstrate that the state has adequate legislative authority to implement the federal CAA programs.

Provisions for Revision Since the federal requirements constantly change, the state SIP must allow for modification as necessary to account for changes in NAAQS and plans to meet the new NAAQS. The fundamental purpose of the SIP is to achieve and maintain compliance with the NAAQS. If the NAAQS change, the SIP must be adjusted to meet the new standards.

Ambient Air Quality Modeling Ambient air quality dispersion modeling is used to "predict" the ambient air quality impact of a proposed new source or modification. Dispersion modeling is fraught with uncertainties, and there have been numerous arguments regarding models and their use (see Chapter 4). However, the federal CAA [Section 110(a)(2)(K)(i)] specifically requires the states to allow modeling for the purpose of predicting the impacts of sources.

Fees The 1990 federal CAA contained a major new emission fee concept that will affect nearly every existing and new source. These emission fees are part of the National Operating Permit Program (see Chapter 10) and are required to be sufficient to cover the costs of implementing the program. However, the CAA also requires in Section 110(a)(2)(L) that the SIP contain fee collection provisions sufficient to cover the costs of reviewing and acting on permit applications submitted by individual facilities and the reasonable costs of implementing and enforcing provisions of the permit. The states each have considerably different fee programs.

TRIBAL IMPLEMENTATION PLANS

Sections 110 and 302 of the Act specifically authorize the EPA to treat Indian tribes as states with respect to implementing the CAA on their lands. The Act also allows the EPA to provide financial assistance to the tribes to implement the Act. In order for the tribe to have primacy, however, the tribe must promulgate and obtain approval for a Tribal Implementation Plan (TIP). In addition to requirements for an approvable SIP, the tribes must also, per Section 301(d)(2) of the CAA,

1. Have a governing body capable of carrying out substantial governmental duties
2. Give the governing body the ability to exercise functions pertaining to the management and protection of air resources (e.g., establishment of a Tribal Environmental Protection Agency)
3. Have the capability and resources necessary to carry out the functions required by the Act

If a tribe has not applied for or received approval of its TIP, the EPA (*not* the states) administers the Act on tribal lands. This can lead to significant confusion regarding applicable regulations, as the following example demonstrates.

One of the keys to determining applicability of various regulatory programs in New Source Review (see Chapter 8) is the potential to emit certain levels of pollutants (i.e., thresholds). "Potential to emit" is defined as the emissions from a device at full capacity taking into account federally enforceable limitations (e.g., control systems) required to be placed on the device. Essentially, this means that the potential to emit is determined after applicable controls have been installed. However, the problem becomes what is a federally enforceable limitation.

Take the case of a proposed new device (source) to be located on state lands (not Indian lands). If the state-approved SIP requires the source owner to install a control system that removes 98 percent of a specific pollutant, the potential to emit would be only 2 percent of the uncontrolled emissions. In this case, the control device would be considered a federally enforceable limitation as it is part of the SIP. On the other hand, if the same source were to be located on Indian land (even if the Indian land was in the same state), and there was no EPA requirement that the device have a 98 percent removal system, the potential to emit would be the full uncontrolled emissions, even if the source owner voluntarily offered to install the control system. This is due to the fact that on the Indian land, the SIP is not applicable, and in the absence of an approved TIP, only compliance with federal regulations is required. As is clear, the tribal-versus-state-lands-and-regulations issue leads to numerous complications.

SANCTIONS AND FEDERAL IMPLEMENTATION PLANS

Although the CAA relies on the principle that state and local governments are responsible for implementing the Act, the Act also provides for penalties to the states if they do not write, obtain approval for, or implement a SIP consistent with federal requirements. These penalties are termed *sanctions* and are detailed in Section 179 of the Act.

One of the driving forces for the 1990 Amendments was that the EPA was being forced to apply sanctions to numerous U.S. cities that had not come into attainment per the deadlines established by the 1977 Amendments. These sanctions include:

1. Withholding federal highway funds (except for safety-related and emission reducing—e.g., high-occupancy vehicle lanes efforts)
2. Withholding EPA grants for state air pollution planning and control programs
3. An increase in the offset ratio (see Chapter 9) to a ratio of 2 to 1

In addition to the sanctions, the EPA is required to implement the CAA at the local level. This implementation is termed a Federal Implementation Plan (FIP) and will include the same steps, or more stringent steps if desired, than the states would have taken to bring the areas into compliance. Obviously, no one wants to have an area come under regulation through a FIP. Therefore, the urgency for a state to develop and implement approvable regulations at the state and local level is very great.

PROCESS OF DEVELOPING REGULATIONS

As indicated, regulations included as part of a SIP cannot be promulgated unless the public has had an opportunity to participate in the development. This process is often time-consuming and frustrating to an environmental manager, because participating in all of the various federal, state, and local regulatory development processes is often not feasible for an environmental manager. In addition, it is often thought that participation by a single facility manager will probably not be successful. However, this way of thinking is incorrect. Participation in the regulatory development process is critical and can be successful. The key to success is knowing when and how to participate.

Regulatory Development at the Federal Level

On a federal level, the regulatory process proceeds through most of the following steps:

1. Regulatory agenda
2. Advanced notice of proposed rulemaking (ANPR)
3. Comment on ANPR
4. Working groups and regulatory negotiation
5. Proposed rule
6. Comment on proposed rule
7. Final rule
8. Litigation
9. Revised final rule in response to litigation

The EPA semiannually (in April and October) publishes a regulatory agenda that summarizes current and projected rulemakings, reviews of existing regulations, and significant regulatory actions taken in the previous six months. This regulatory agenda is the first official notice that EPA is about to promulgate a rule. Along with the agenda, an example of which is shown in Figure 3-2, the EPA lists a contact person for each action. If the environmental manager is concerned about a future rule, establishing contact with the designated EPA staff person at this stage can be very useful. The EPA staff person wants and needs input as early in the process as possible, and establishing a relationship at this time will provide great benefits later in the regulatory development process.

Often, the EPA does not meet the schedules established in the regulatory agenda, and a rule can remain in limbo for a very long time. Actual activ-

3766.NESHAP:PRINTING/ PUBLISHING INDUSTRY

Legal Authority: Clean Air Act Amendments of 1990, sec 112

CFR Citation: 40 CFR 63

Legal Deadline: Final, Statutory, November 15, 1994
EPA is required to promulgate 25% of the listed source categories by November 15, 1994. EPA plans to promulgate this action by that date to satisfy part of this statutory obligation.

Abstract: This regulation would control hazardous air pollutant emissions from printing/publishing facilities. The Agency is currently gathering background information on the rotogravure, flexography, offset lithography, screen printing, letterpress and other parts of the printing/publishing industry. The Agency will make a decision on the scope of this regulation once this work is completed.

Timetable:

Action	Date	FR Cite
NPRM	11/00/93	
Final Action	11/00/94	

Small Entities Affected: Undetermined

Government Levels Affected: Undetermined

Additional Information: SAN No. 3167.

Agency Contact: David Salman, Environmental Protection Agency, Air and Radiation, (MD-13), Research Triangle Park, NC 27711, **919 541-0859**

RIN: 2060-AD95

3767.NESHAP:POLYMERS AND RESINS, GROUP I

Legal Authority: Clean Air Act Amendments of 1990, sec 112

CFR Citation: 40 CFR 63

Legal Deadline: Final, Statutory, November 15, 1994
EPA is required to promulgate 25% of the listed source categories by November 15, 1994. EPA plans to promulgate this action by that date to satisfy part of this statutory obligation.

Abstract: This regulation under development would control emissions of hazardous air pollutants (HAPs) from the manufacture of the butyl rubber, epicholohydrin elastomer, ethylene propylene rubber, Hypalon (TM), neoprene, nitrite butadiene rubber, polybutadiene rubber, polysulfide rubber and styrene butadiene rubber and latex. Emissions from process vents, equipment leaks, wastewater and storage will be addressed by this regulation for both new and existing sources.

Timetable:

Action	Date	FR Cite
NPRM	01/00/94	
Final Action	01/00/95	

Small Entities Affected: Undetermined

Government Levels Affected: None

Additional Information: SAN No. 3166.

Agency Contact: Leslie Evans, Environmental Protection Agency, Air and Radiation, (MD-13), Research Triangle Park, NC 27711, **919 541-5410**

RIN: 2060-AD96

■ **FIGURE 3-2 Example of an EPA regulatory agenda.** *Source:* Federal Register, 1993.

ity on a new rule begins with promulgation of an Advanced Notice of Proposed Rulemaking (ANPR). The ANPR states the legislative and regulatory basis for the proposed rule, lists the options being considered by the EPA for the rule, and requests comments from interested parties. The ANPR may also request and establish a working group or regulatory negotiation committee and may announce various public meetings at which the rule and options for implementing the goals of the rule will be discussed.

Interested managers should respond and provide comments at this stage. The comments should be as specific as possible and deal with implementing the goal of the rule rather than challenging the entire purpose of the rule. Generally, the EPA is required by legislative directive (i.e., the U.S. Congress) to promulgate the rule. Therefore, challenging the necessity of the rule is not very fruitful. At this stage, comments will generally focus on the technical issues that must be resolved and provide suggested conceptual methods for resolving those issues. This type of comment and issue resolution continues through the fourth step of the process: regulatory negotiation and public workshop committees.

The fifth step of the process is publication of a Proposed Rule. The proposal is very detailed and includes responses to comments that the Agency received on the ANPR, lists the reasons for the Proposed Rule, and suggests regulatory language. The Proposed Rule Preamble will also often request additional comments on specific technical issues that have not yet been resolved.

Some managers may feel that once a Proposed Rule is published, there is no further benefit in commenting on the rule. On the contrary, this author feels that commenting on the Proposed Rule is crucial to proper interpretation and implementation of the Final Rule. When the Final Rule is published, the Agency must respond to each of the meaningful comments and describe how the Final Rule addresses those comments.

An example of the importance of commenting on proposed rules is related to the Prevention of Significant Deterioration (PSD) regulations (see Chapter 8). The definition of "source" is very important in PSD regulation and includes a provision that the emitting devices must be located on adjacent or contiguous property. "Contiguous" is defined as two parcels of property in close proximity, but not sharing a common boundary. (Adjacent property includes two parcels sharing a common boundary.) The separation distance at which properties are no longer contiguous is not defined in the PSD regulations. However, during the Proposed Rule stage (in 1978) of the PSD regulations, one commenter asked the EPA whether two devices located 26 miles apart would be considered contiguous. The EPA responded in the Preamble to the Final Rule that 26 miles was "too far" to be considered contiguous.

Unfortunately, a major oil and gas exploration and production company had facilities located 18 to 22 miles apart, and the local regulatory agency wanted to consider the facilities as being contiguous for purposes of New Source Review (see Chapter 8). Because the only definitive comment the EPA had made regarding this issue indicated that 26 miles was "too far," the local agency prevailed in considering 18 to 22 miles as being contiguous. It is almost certain that if the original commenter had asked whether 18 miles was too far to be contiguous, the EPA would have agreed when the rules were initially promulgated.

This example shows how direct, specific comments can have very far-reaching and long-range impacts on how the rules are interpreted and implemented. Therefore, even though comments may not change a rule, they certainly can have a bearing on the subtleties of interpretation contained within the rule.

All of the technical research and backup for proposed and final Agency action, comments from the public, and responses to comments are contained in the federal docket for the rule. The docket can be accessed by the general public and contains much useful information (as well as much useless information). If one wants to fully understand all aspects of a regulation and/or litigate the Final Rule, the federal docket contains information that must be accessed. The Federal Register gives a docket number and location for each agency action.

After a Final Rule is promulgated, there is nearly always litigation associated with it. The litigation revolves around those technical and regulatory aspects of the rule that were not addressed by EPA when the rule became final. Unless a party obtains an injunction preventing enforcement of a rule, the published Final Rule must be complied with until a court reverses the Agency.

The courts, in general, will defer to the Agency for technical evaluations and decisions. Therefore, litigation is most often successful when it can be shown that there were procedural errors in promulgation of the rule, or the Agency did not address substantive comments or concerns. From a practical view, it is this author's opinion that litigation only complicates the regulatory process; it rarely makes the rules significantly more or less stringent.

If the Agency loses the litigation, it will be required by the court to reissue the rule and/or modify it in accordance with the court's instructions.

Often, the Agency will be instructed to reexamine the rule and evaluate whether or not it should be changed to address the plaintiff's concerns. This is called "remanding" the rule. Practically speaking, what the agency does is to promulgate an interim rule just like the old rule while the remand investigation is ongoing. This proves troublesome for an industry environmental manager, because it is not known whether the revised rule will be more or less stringent than the original rule promulgated prior to the remand. A good practice is to assume that the Agency will repromulgate the rule at least as stringently as the original rule (i.e., ignore the litigation).

Regulatory Development at the State and Local Level

The state and local regulatory development process is analogous to the federal. At the state and local level, however, there seems to be more emphasis on offering regulatory development workshops where the local agencies work more closely with the regulated community and members of the concerned public to achieve a workable regulation. The workshop process can take a very long time at the state and local level, but, again, it is critical to participate in the workshop process. If one waits until a rule is publicly proposed, it is very difficult for an industry to be able to change it.

Principles for Effective Regulatory Process

Working with the regulatory agencies requires strategy and professionalism. The mistakes made when dealing with an agency include not being prepared, making comments that are too general, and asking questions before thinking through the possible answers. The following principles for an effective regulatory process are useful to keep in mind when dealing with an agency at any stage of the regulatory process, from proposed rule to obtaining permits. The principles are as follows (Erbes, 1993):

1. *Do your homework.* Know the regulations, legislation, guidance documents, and actual physical impacts of your operation. Know the technical logic behind the regulations.
2. *Be consistent.* Change technical and regulatory approaches, design, and operating characteristics for only known and communicated reasons.
3. *Be proactive.* Anticipate the regulatory agency's questions and concerns. Provide the answers before being asked. Do not leave issues unresolved.
4. *Have a plan.* Implement the plan and meet your commitments. Squarely meet the regulatory needs and cut up big, complex projects into manageable units. Establish goals for each regulatory process.

5. *Do not attempt the impossible.* Do not accept impossible requests by the agency, even for expediency.
6. *Do the right thing.* If possible, do not accept scientifically or regulatorially incorrect demands, even for expediency.
7. *Ask confirming questions only.* Do not ask a regulatory agency a question unless you have done your homework, and you think you know the answer, or unless you have decided that open-ended questions are in your best interest.
8. *Tell the agency what is happening.* Do not keep the agency in the dark. Give complete information relevant to the environmental issue, and avoid providing speculative and irrelevant information.
9. *Recognize that it is your project, not the agency's.* Manage the agency process just as you manage your department. Similarly, help the agency by anticipating and resolving agency issues.
10. *Be professional.* Develop a good-faith, personal, and cooperative working relationship with the agency people responsible for your project.

DIFFERENCES BETWEEN PERMITS AND REGULATIONS

One of the main sources of confusion in air quality compliance on a daily operational basis is the difference between permits and regulations promulgated at the state and/or local level. A permit is issued for a specific device at a specific facility by name and address. On the other hand, regulations provide overall minimum requirements that all devices must meet.

The difficulty comes in that, at times, the permit is not consistent with regulations, either because when the permit was issued a regulation was misinterpreted, or a regulation changed after a permit was issued. However, it is critical to note that *both* the permit and regulations must be complied with. For major sources (termed Title V or "Part 70" sources), the new national operating permit program will provide a "permit shield" that provides some protection against this problem (see Chapter 10). However, for minor sources that do not require Title V permits, the differences between regulation and permits remain important.

The following example illustrates the differences between permits and regulation.

In this example, a facility had a permit for a paint booth that limited the total use of coatings to 3 gallons per day. A copy of the permit is shown in Figure 3.3.

CLEAN AIR QUALITY MANAGEMENT DISTRICT

PERMIT TO OPERATE

M98765

1234 CLEAR AIR LANE, CLEAR VALLEY, CALIFORNIA 12345

Operation under this permit must be conducted in compliance with all information included with the initial application and the initial permit conditions. The equipment must be properly maintained and kept in good operating condition at all times. In accordance with Rule ABC, this Permit to Operate or copy must be posted on or within 8 meters of equipment.

LEGAL OWNER OR OPERATOR: PAINT INDUSTRIES APPL. NO. 12345

EQUIPMENT LOCATED AT: 12345 S. MAIN STREET
CLEAR AIR CITY, CALIFORNIA

EQUIPMENT DESCRIPTION AND CONDITIONS:

SPRAY BOOTH, BLEEKER, FLOOR TYPE, 6'-4" W. X 6'-2" L. X 7'-0" H., WITH TWELVE 20" X 20" EXHAUST FILTERS AND ONE 1-1/2 H.P. EXHAUST FAN.

PAGE 1 OF 2

This initial permit must be renewed by 08/16 ANNUALLY unless the equipment is moved, or changes in ownership. If billing for annual renewal fee (Rule ABC) not received by expiration date, contact above office.

This permit does not authorize the emission of air contaminants in excess of those allowed by Division 26 of the Health and Safety Code of the State of California or the Rules of the Air Quality Management District. This permit cannot be considered as permission to violate existing laws, ordinances, regulations or statutes of other government agencies.

CLEAN AIR QUALITY MANAGEMENT DISTRICT

CONTINUATION OF PERMIT NO. M98765 APPL. NO. 12345

— CONDITIONS —

1. THIS SPRAY BOOTH MUST NOT BE OPERATED UNLESS ALL EXHAUST AIR PASSES THROUGH FILTER MEDIA AT LEAST 2 INCHES THICK.

2. A GAUGE MUST BE INSTALLED TO INDICATE, IN INCHES OF WATER, THE STATIC PRESSURE DIFFERENTIAL ACROSS THE EXHAUST FILTERS. IN OPERATION, THE PRESSURE DIFFERENTIAL MUST NOT EXCEED 0.25 INCH OF WATER.

3. THE TOTAL QUANTITY OF COATINGS AND SOLVENTS USED IN THIS EQUIPMENT MUST NOT EXCEED 3 GALLONS IN ANY ONE DAY.

4. THIS EQUIPMENT MUST BE OPERATED IN COMPLIANCE WITH RULE ABCD.

5. THE OPERATOR SHALL KEEP ADEQUATE RECORDS TO VERIFY DAILY USAGE. SUCH RECORDS SHALL BE RETAINED FOR A PERIOD OF TWO YEARS.

■ **FIGURE 3-3** **Example of a paint spray booth permit.**

The permit was issued on March 25, 1987, and the 3-gallon-per-day limit was consistent with the limit previously established for the paint booth drying oven in a permit issued on October 1, 1982. At the time that the paint booth permit was previously issued, there was no regulation in this jurisdiction that proscribed any other limits on the paint booth other than records of daily coating usage, which were required to be kept for a period of five years.

The facility dutifully kept its records, and the average recorded coating use was recorded as 1.15 gallons per day, although *actual* usage was suspected to be about 0.8 to 0.9 gallons per day. The reason for the difference was that, because the limit was 3 gallons and the usage was well below the limit, the paint booth operator tended to approximate and overestimate coating usage.

However, nine months after the permit was issued, a new regulation was promulgated that limited the usage of a certain type ("clear coat") of coating to less than 1 gallon a day unless the paint booth had an emission control system. Unfortunately for this specific facility, it was using that specific type of clear coat that was limited by regulation to 1 gallon per day. Even more unfortunately, the facility was not aware of this change in regulation and continued to think that only the permit 3-gallon-per-day limit applied.

During an inspection by the local air quality agency, it was found that the facility had been recording the use of more than 1 gallon per day of clear coat. Accordingly, the agency fined the facility $250,000 for 250 days of exceeding the 1-gallon limit and violating the regulation, even though the facility was operating within its permit. The facility was able to negotiate a lesser fine, but the facility still expended a considerable amount of resources and money to defend itself. The mistake was to presume that compliance with the permit was sufficient for compliance with regulations.

It is important for environmental managers to actively track and evaluate new regulatory developments. One cannot rely only on mailing lists and existing, outdated permits. Personal contact with the Agency on a regular basis is important.

CHAPTER

4

AMBIENT AIR QUALITY STANDARDS AND ATMOSPHERIC DISPERSION

Traditionally, at the federal level, the quality of the air we live in is measured by comparing pollutant concentrations to the National Ambient Air Quality Standards (NAAQS). At the state and local level, it is measured by comparing the pollutant to NAAQS as well as State Ambient Air Quality Standards (SAAQS) and local ambient air quality standards, if any. As will be seen in Chapters 5 and 8, recent legislation (1977 and 1990 CAA Amendments) has added other parameters such as aesthetics (visibility) and risk (air toxics) to the valuation of air quality.

FEDERAL AMBIENT AIR QUALITY STANDARDS

National ambient air quality standards are promulgated by the EPA in order to protect the public health and welfare with an adequate margin of safety. The EPA has promulgated NAAQS for six pollutants: sulfur dioxide (SO_2), nitrogen dioxide (NO_2), carbon monoxide (CO), lead (Pb), ozone (O_3), and particulates (PM_{10}). These pollutants are termed the "criteria" pollutants or "traditional" pollutants.

Definition of Ambient Air

The NAAQS apply to ambient air, which is air in the breathing zone to which the public has access. Therefore, the NAAQS apply to the air located approximately 1.5 meters (about 5 feet) above ground level and at locations beyond the controlled perimeter ("fenceline") of a facility. If the facility has a well-controlled and obvious fence surrounding its property that restricts public access, the location of ambient air is relatively straightforward. Even if the facility has a parking lot within the fence, since the parking lot is for employees of the facility, the NAAQS apply at and beyond the fenceline, not in the parking lot. [There may be other regulations (e.g., OSHA) that set standards for nonpublic areas.]

However, there are several cases when it is not obvious where the NAAQS should be applied. For example, a college campus is owned and controlled by the college, but the general public has free access throughout the campus. Therefore, it could be argued that the NAAQS apply everywhere on the campus except inside buildings. Another example is hospitals. Hospital grounds are controlled, and in many cases the grounds are even fenced (especially in the case of military and veterans hospitals). However, since the public can enter the grounds without registration or obstruction, the NAAQS probably apply throughout the grounds.

Military installations (i.e., naval bases, naval air stations, air force bases, army posts, etc.) also pose a special problem. This is because, even though the perimeter of the installation is fenced and access completely controlled (by requiring registration at the guard gate), members of the public can enter the base given sufficient reasons. Some military installations even have public museums and other facilities on the installation that encourage public visitors. However, in general, regulatory agencies have considered the NAAQS to apply only at the perimeter of the installation and beyond. (As discussed in Chapter 8, this decision can be a double-edged sword.)

As is explained below, a change in location where the NAAQS apply changes the predicted impact (through atmospheric dispersion modeling) of a source of pollution and can affect whether or not the source can obtain a permit. Therefore, the question "What is ambient air?" is not trivial.

Definition of Atmosphere

Air pollutants are emitted into the atmosphere. If a chemical is not emitted into the atmosphere, it is not considered an air pollutant. Generally, the air within a building is not considered the atmosphere for determining whether an air pollutant has been emitted. (The air within the workplace is protected through OSHA regulations and not through air quality regulations.)

However, some local jurisdictions define "atmosphere" to include the air within a building, and some jurisdictions define atmosphere both ways. For example, in the Los Angeles area, the local permitting authority, the South Coast Air Quality Management District (SCAQMD), defines atmosphere to include the air within a building in the Los Angeles Basin (where pollution is severe). However, in the high desert area of the SCAQMD, where pollution is not as severe, "atmosphere" refers only to air outside the building.

Criteria Pollutants

As indicated, the NAAQS have been promulgated for six criteria pollutants. These pollutants are termed "criteria pollutants" because the CAA requires the EPA to evaluate the potential health effects of these six pollutants and to issue a criteria document once every five years. The criteria documents detail the health studies and other scientific information evaluated in the process of setting the NAAQS at a level that protects the public health and welfare with an adequate margin of safety.

Although the Act requires a new criteria document once every five years, EPA usually takes about 10 years or more to issue new criteria documents and consider revising the standards. The most recent change in standards occurred in 1987 when the particulate standard was changed from total suspended particulate (TSP) to particulate with a mean aerodynamic diameter of 10 microns or less (PM_{10}). (The difference will be discussed later in this chapter.) Since that time, the EPA has reviewed the nitrogen dioxide standard, the ozone standard, and the sulfur dioxide standard and has chosen to leave the standards as originally promulgated. The standards continue to be reviewed and the EPA is considering adding a $PM_{2.5}$ particulate standard and a short-term (5-minute to 1-hour) SO_2 standard, among other changes.

Contrary to the participative effort recommended in previous chapters for affecting the development of proposed and final rules and regulations, participation by the normal environmental manager in the criteria document process is not very beneficial. This is because the NAAQS must be set purely for health and welfare reasons. The ability of an industry to meet the standards cannot be considered. Therefore, only health effects data and information are relevant to the NAAQS setting process.

Federal NAAQS

Table 4-1 shows the federal NAAQS. Note that the NAAQS are written for different averaging times, and compliance with the NAAQS is different for the different pollutants.

The reasons for the differences are related to the criteria documents. For example, for sulfur dioxide, the conclusion of EPA, based upon the criteria document, was that three different averaging times would be needed in order to protect the public health and welfare with an adequate margin of safety.

Primary standards are those needed to protect public health, whereas secondary standards are those needed to protect the public welfare. The criteria document for SO_2 indicates that, with respect to health, there are both daily and annual SO_2 exposure concerns. With respect to welfare, only a 3-hour exposure is of concern. (The Agency uses known meteorological variability

■ TABLE 4-1 National Ambient Air Quality Standards[a]

Pollutant	NAAQS
Sulfur oxides (sulfur dioxide)	
Annual	80 (0.03 ppm)
24 hours	365 (0.14 ppm)[b]
3 hours	1300 (0.5 ppm)[b]
Particulate (PM_{10})	
Annual	50[c]
24 hours	150[d]
Carbon monoxide	
8 hours	9 ppm (10,000)[b]
1 hour	35 ppm (40,000)[b]
Ozone	
1 hour	0.12 ppm (235)[d]
Nitrogen dioxide	
Annual	0.053 ppm (100)
Lead	
Quarterly	1.5

[a]Concentrations in μg/m³ and arithmetic average unless noted. Conversions at 760 mm Hg, 25°C.

[b]Not to be exceeded more than once per year.

[c]Statistical expected value.

[d]Statistical expected exceedance one day or less per year.

and relationships between various averaging times to ensure that all possible exposure durations are evaluated and covered by the standard.) Therefore, for SO_2 there is a 3-hour, 24-hour, and annual standard. Also note that these values can be exceeded in an area once (but no more than once) a year and the area is still considered to meet the standard (termed "attainment").

Although there is a distinction between the primary and secondary standards, from a practical compliance perspective, the distinction is no longer important. (It was relevant for the attainment deadlines contained in the 1977 CAA, but these deadlines have been replaced by the 1990 CAA. See Chapter 9.)

Standards Averaging Times

The NAAQS averaging times are an important practical compliance issue, because when siting a new source and evaluating the compliance of the new source with respect to the NAAQS, the evaluation is made on a specific pollutant and averaging time basis. As discussed in the following paragraphs regarding ambient air quality modeling, ambient impact is related to pollutant emissions through atmospheric dispersion. Although ambient standards are stated in terms of various averaging times, emissions are usually expressed in terms of an hourly emission rate (e.g., pounds per hour) or in terms of parameters (grains per dry standard cubic foot) measured over an hourly period.

The standard regulatory practice is to assume that the emitting device emits the maximum pounds per hour emission rate continuously (i.e., 365 days per year, 24 hours per day, for a total of 8,760 hours per year). However, no device operates continuously at full capacity for an entire year (there is, at minimum, down time for maintenance). If dispersion modeling indicates that the annual standard for a pollutant may be exceeded, the device operator may wish to choose to accept a permit limitation prohibiting the device from operating at the theoretical maximum capacity. Such limitations must be "enforceable," and the agencies do not generally like to accept such limitations. However, they are feasible.

Therefore, when specifying equipment emission rates and/or obtaining vendor quotes for emission rates, it is critical to specify not only the pollutant but also the averaging time. Emission rates for all of the NAAQS averaging times should be obtained for each pollutant emitted.

Chemical Differences in Pollutants

There are several technical subtleties contained within the NAAQS and SAAQS. These subtleties relate to

1. The specific chemicals for which the standards are promulgated
2. The physical form of the pollutant, for which the standards are promulgated
3. The method of evaluating compliance with the standard

Sulfur Oxides By regulation (40 CFR 50.4), the NAAQS for sulfur oxides include all oxides of sulfur, measured as sulfur dioxide. Sulfur dioxide is the normal oxide found in ambient air, and the question of sulfur oxides versus sulfur dioxide is not normally an issue. The situation is quite different for nitrogen dioxide, however.

Nitrogen Oxides One of the differences is related to oxides of nitrogen (NO_x) versus nitrogen dioxide (NO_2). Note that the ambient standard is for NO_2. However, equipment vendors and source emissions tests usually report emissions in terms of NO_x. Combustion sources generally emit mostly nitrogen monoxide (NO). But NO is not stable in the atmosphere and eventually converts to NO_2. Therefore, regulatory agencies usually assume that all NO_x is NO_2 when conducting impact modeling. On the federal level, this assumption may be reasonable, because there is only an annual average NAAQS, and over the period of a year there is time for the NO to convert to NO_2. However, if the maximum impact point of the source is very near the source, there may not be enough time for the NO to react to form NO_2 by the time the NO reaches the impact point. Therefore, if a source emits mostly NO (and not NO_2), there may be some adjustment allowed by the agency for unreacted NO in the impact modeling.

The difference between NO_x and NO_2 is most important on a state level. Many states have a 24-hour or shorter NO_2 state standard. In this case, not all NO emitted will react to form NO_2 within a 24-hour period, and some adjustment to the assumption regarding "NO equals NO_2" is required. One of the ways to address this problem is to use the Ozone Limiting Method (OLM) model in which the fraction of NO converted to NO_2 is a function of the ambient concentration of ozone in the atmosphere and the time required for transport of the pollutant from the emitting point to the ambient receptor. (For example, if the wind speed is 2 meters per second and the property boundary is 300 meters away from the emitting point, the transport time is 150 seconds and in that short of a time, not much of the NO will have reacted to form NO_2.) Therefore, the concentration of NO_2 will be less. A second method is the Ambient Ratio Method (ARM).

Another issue with respect to NO_x is that dinitrogen oxide (N_2O) is also an oxide of nitrogen and a common gas. This gas is used as a mild anesthetic ("laughing gas") and is relatively stable in the atmosphere. Even though N_2O can be considered part of NO_x, it is not a regulated pollutant and is not evaluated with respect to the NAAQS.

Ozone Facilities do not generally emit ozone. The ozone standard is met by regulating gases that photochemically react in the lower atmosphere (the troposphere) to create ozone. (Note that the CAA attempts to address both tropospheric ozone, where there is too much, and stratospheric ozone, where there is not enough.) Several different terms are used to describe these reactive gases, but the terms are not synonymous. The terms often used are

1. Total nonmethane hydrocarbons (TNMHCs)
2. Nonmethane organic gases (NMOGs)
3. Volatile organic compounds (VOCs)
4. Reactive organic gases (ROGs) or reactive organic compounds (ROCs)

The most precise and *technically* correct term is ROG. However, the CAA uses the term VOC, and VOC is the proper *regulatory* term. Generally, ROG and VOC are synonymous. VOC is defined by regulation [40 CFR 51.100(aa)] to include any gaseous chemical compound that contains the element carbon and that participates in atmospheric photochemical reactions, except for excluded compounds. The excluded compounds are carbon monoxide, carbon dioxide, carbonic acid, carbonates, ammonium carbonate, metallic carbides, and the following chemicals considered by the EPA to have negligible photochemical reactivity:

1. Methane
2. Ethane
3. 1,1,1-Trichloroethane
4. Methylene chloride
5. CFC-11, 12, 23, 113, 114, 115
6. FC-23
7. HCFC-123, 124, 125, 134, 134a, 141a, 142b, 152a
8. Perfluorocarbon compounds with
   a. Cyclic, branched or linear, completely fluorinated alkanes
   b. Cyclic, branched or linear, completely fluorinated ethers with no unsaturations

   c. Cyclic, branched or linear, completely fluorinated tertiary amines with no unsaturations
   d. Sulfur containing perfluorocarbons with no unsaturations and with sulfur bonds only to carbon and fluorine

9. Acetone

It is important to note, however, that many local jurisdictions have a slightly different definition of VOC, and some of the exempt compounds may become listed in later updates to even the federal regulations. Usually, at the local level, the number of exempt compounds is reduced compared to the federal (i.e., more compounds are included in the definition of VOC at the local level).

The VOC regulation is designed to include all of the volatile organic compounds that are photochemically reactive in the lower atmosphere. However, note that although 1,1,1-trichloroethane (TCA) is exempt as a VOC, it is an ozone-depleting compound (ODC) and is regulated by the CAA (Chapter 12). Methylene chloride is a suspected carcinogen and it, too, is regulated under both the HAPs program (Chapter 7) and as a local air toxic under various local regulatory programs (see Chapter 5).

Chemically, VOC is a subset of NMOG, and NMOG is a subset of TNMHC. Obviously, NMOG includes all gases other than methane, and TNHMC includes all gases and nongases (e.g., particles) other than methane. The reason the distinctions are important is that equipment vendors and source emissions testing firms often report VOC emissions non-speciated (i.e., without distinguishing the specific chemicals) and often report NMOG rather than VOC. Therefore, there most often is an overestimate of the "regulatory" emissions (i.e., VOC) when using vendor emission rates or source emission tests that quote NMOG. Likewise, TNMHC is clearly not an appropriate parameter since it includes particles that are not reactive in the atmosphere (such as polyaromatic hydrocarbons, or PAHs).

Physical Differences in Pollutants

Physical differences in pollutants are as important as chemical differences. The particulate matter NAAQS is in terms of PM_{10}. PM_{10} is defined as those particles whose mean aerodynamic diameter is less than or equal to 10 microns. Traditionally, particulate matter was regulated as total suspended particulate (TSP). TSP was defined as those particles collected by an EPA reference standard high volume (hi-vol) sampler and is now generally considered to include particles up to about 55 microns in diameter.

TSP was the first pollutant regulated by the EPA. Therefore, equipment vendors and source emission test results are most often in terms of TSP. However, it is clear that PM_{10} is a subset of TSP, and TSP is an overestimate of PM_{10}. EPA changed the NAAQS to reflect only PM_{10} in 1987, but they did not change the PSD increment (see Chapter 8) to reflect only PM_{10} until May 1993.

When the PM_{10} standard was promulgated, the EPA stated that, on average, PM_{10} is about one-half TSP. This ratio can be used as a rule of thumb when converting TSP data to PM_{10} data. However, the ratio depends on the source of the particulate matter. When the source is a chemical or combustion process, the ratio of PM_{10} to TSP is probably greater than 1 to 1 (i.e., the PM_{10} percentage is greater than 50 percent). However, when the source of particulate is a physical process (e.g., dust generated by soil handling or vehicular traffic), the ratio of PM_{10} to TSP is much less than 1 to 1 (i.e., the PM_{10} percentage is significantly less than 50 percent). This is due to the fact that physical processes usually do not create the smaller particles, because the electrostatic forces between particles tend to keep the particles together and, therefore, the particles retain their larger diameter.

Another subset of particulate matter is $PM_{2.5}$. These are particles with mean aerodynamic diameters of less than 2.5 microns. The EPA is considering adding a $PM_{2.5}$ standard primarily to protect visibility.

Pollutant Precursors

When assessing the ambient air quality impact of a source, one must be careful to assess not only directly emitted pollutants, but pollutant precursors as well. A precursor is a substance that, when released to the atmosphere, undergoes chemical reaction or transformation such that the reaction product is a pollutant for which NAAQS or SAAQS have been established. For example, when ammonia is emitted into the atmosphere, if sulfur dioxide is present, the ammonia will react with the sulfur dioxide to form an ammonium sulfate salt, which is a particulate. Therefore, ammonia could be considered a precursor to particulates.

The following are examples of precursors and their secondary pollutants (i.e., reaction products):

Precursors	Secondary Pollutants
Reactive organic gases	Photochemical ozone Organic particulate
Nitrogen oxides	Nitrogen dioxide Nitrate particulate Photochemical ozone
Sulfur oxides	Sulfur dioxide Sulfates Sulfate particulate

Physical Reference Conditions

The NAAQS are promulgated in terms of volume of pollutant per volume of air (i.e., parts per million). (The precise abbreviation for parts per million on a volume basis is ppm(v); however, common practice in air quality uses the abbreviation ppm.) This is consistent with how most of the ambient monitoring instrumentation provides results for the gaseous pollutants. However, ambient dispersion models yield results in terms of mass per volume (e.g., $\mu g/m^3$). In addition, some SAAQS are stated in terms of mass per volume instead of volume per volume.

Because the volume of a pollutant is a function of the temperature and pressure at the observation point, a conversion convention must be defined. The federal regulations (40 CFR 50.3) state that all standards are referenced to a pressure of 760 millimeters of mercury (i.e., sea level) and a temperature of 25 degrees Celsius. At higher elevations (more than a few thousand feet above mean sea level) and higher temperatures, the difference between the volume-per-volume measurement and the converted mass-per-volume standard can be significant, on the order of 10 percent to 30 percent. Therefore, this difference must be kept in mind when evaluating and reporting monitored data.

STATE AND LOCAL AMBIENT AIR QUALITY STANDARDS

The states often promulgate ambient air quality standards that are more stringent than the federal NAAQS. The states also often include additional pollutants and averaging times. For illustrative purposes, Table 4-2 shows the California State standards and Table 4-3 shows the Washington State standards.

TABLE 4-2 California State Ambient Air Quality Standards[a]

Pollutant	Standards
Sulfur oxides (sulfur dioxide)	
Annual	—
24 hours	0.04 ppm (105)
3 hours	—
1 hour	0.25 ppm (665)
Particulate (PM_{10})	
Annual	30
24 hours	50
Particulate (TSP)	
Annual	—
24 hours	—
Carbon monoxide	
8 hours	9 ppm (10,000)
1 hour	20 ppm (23,000)
8 hours (Tahoe)	6 ppm (6,670)
Ozone	
1 hour	0.09 ppm (180)
Nitrogen dioxide	
Annual	—
1 hour	0.25 ppm (470)
Lead	
Quarterly average	1.5
Sulfates	
24 hours	25
Hydrogen sulfide	
1 hour	0.03 ppm (42)
Vinyl chloride	
24 hours	0.010 ppm (26)
Visibility	
B_{ext}	0.23 km^{-1}
B_{ext} (Tahoe)	0.07 km^{-1}

[a]Concentrations in $\mu g/m^3$ unless noted. Conversions at 760 mm Hg, 25°C.

TABLE 4-3 Washington State Ambient Air Quality Standards[a]

Pollutant	Standards
Sulfur oxides (Sulfur dioxide)	
Annual	0.02 ppm (53)
24 hours	0.10 ppm (261)
3 hours	—
1 hour	0.40 ppm (1040)
Particulate (PM_{10})	
Annual	50
24 hours	150
Particulate (TSP)	
Annual	60
24 hours	150
Carbon monoxide	
8 hours	9 ppm (10,000)
1 hour	35 ppm (40,000)
Ozone	
1 hour	0.12 ppm (235)
Nitrogen dioxide	
Annual	0.05 ppm (94)
1 hour	—
Lead	
Quarterly average	1.5

[a]Concentrations in μg/m^3 unless noted. Conversions at 760 mm Hg, 25°C.

As is clear from Table 4-2, the state of California has a much more extensive list of pollutants and standards. It is also clear that California has different standards for different regions of the state. The reason for these local differences is that the different standards address different political concerns. For example, California has established a more stringent visibility standard for the Lake Tahoe area, which is a very scenic area visited by many tourists.

One must also be careful when evaluating SAAQS to ensure that the proper form of the pollutant is considered. For example, Washington regulates both PM_{10} (by incorporating the NAAQS) and TSP (by retaining the former federal TSP standard as a state standard).

One final thing to keep in mind regarding state standards is that the scientific basis for them is usually not as complete and documented as the EPA's basis for the NAAQS. This means that state standards are more amenable to change should a facility have difficulty meeting a state ambient air quality standard. From a practical point of view, it should be considered nearly impossible to change a NAAQS, but feasible to change a SAAQS.

METHODS OF EVALUATING COMPLIANCE WITH NAAQS

As shown in Table 4-1, for averaging times less than quarterly (e.g., 24-hour, 8-hour, 3-hour), the NAAQS regulations allow one, but no more than one, exceedance of the stated value per year. However, the method of determining compliance differs slightly for each pollutant.

Sulfur Oxides and Carbon Monoxide

For sulfur oxides and carbon monoxide, compliance is simply evaluated by observing the number of exceedances in a given year. If the number is one or less, the area is in compliance. If the number of exceedances is greater than 1, the area is not in compliance (termed *nonattainment*).

Nitrogen Dioxides

Compliance with the annual nitrogen dioxide standard is evaluated on a quarterly basis, such that no single quarter has an average concentration (calculated by averaging all of the hourly concentrations in a quarter) greater than the standard of 0.053 ppm.

Lead

Compliance with the lead standard is evaluated on a quarterly basis, such that no single quarter has an average concentration greater than the standard of 1.5 micrograms per cubic meter.

Particulate Matter and Ozone Statistical Evaluations

The PM_{10} and ozone standards state that the *expected* number of exceedances must be less than or equal to 1. This is different from the sulfur oxide and carbon monoxide standards, which state only that the *number* of exceedances must be less than or equal to 1. To determine the "expected number" of exceedances for PM_{10} and ozone, an average annual expected number of exceedances must be evaluated. EPA regulations state that the expected number of exceedances should be calculated for the immediately preceding 3-year period (generally, calendar-year period).

If complete data for the 3-year period are available (e.g., a 24-hour average concentration is calculated from 24 measured 1-hour values and these values are available for all 365 days in each year), evaluating compliance with the standard is relatively simple. For this situation, the actual number of exceedances in each year is simply totaled, and the total is divided by 3. If this value is greater than 1, the area is not in compliance.

For example, assume that an area had 1 exceedance in the first year, no exceedances in the second year, and 2 exceedances in the third year. In this example, the area would still be in compliance since the sum of the 3 years of exceedances is 3, which, when divided by 3, equals 1. If, however, there were 3 exceedances in the third year, the area would not be in compliance since the sum of exceedances is 4, which when divided by 3 equals 1.3, a value greater than 1.

For situations where the data are not complete, however, the EPA details interpretative methods to supplement missing data. The interpretive methods assume that the ratio of observed exceedances to the actual number of observations will apply equally to the number of days for which no observations are available. Such methods have been published for two pollutants, ozone and PM_{10}. The detailed methodologies are contained in the regulations (40 CFR Part 50, Appendices H and K, respectively). The PM_{10} methodology allows for periods in which no data are obtained (missing data) and for unusual events, while the ozone methodology allows for seasonal differences as well as missing data.

Ozone Correction for Missing Data For ozone, the expected number of exceedances for a single year is determined by the following formula:

$$e = v + [(v/n) \times (N - n - z)] \tag{4-1}$$

where

e = expected number of exceedances in a year
v = actual number of observed exceedances
n = number of days actually monitored with valid data
z = number of days assumed to be less than the standard
N = number of days in the year for which monitoring should occur (normally 365 days per year)

In equation 4-1, the value of z is determined by assuming that if ozone data for the day previous to the missing day and the day after the missing day are both less than the standard, it can be assumed that the missing day would have also been less than the standard. The value for N is normally 365 days per year, but states can obtain a waiver from the EPA to avoid having to monitor year-round, because the ozone concentrations are seasonally dependent (lower in winter). Therefore, the value for N can be less than 365.

One can see from equation 4-1 that, if all 365 days are monitored and have valid data, the expected number of exceedances in a year equals the actual number of exceedances. (This is because if all 365 days are monitored, the term $(N - z - n)$ would be equal to 0.) However, if not all days are monitored, the expected number of exceedances is equal to the actual number of exceedances increased by a ratio to account for missing days.

Recall also that the ozone standard is an 8-hour standard. If an 8-hour period is observed to be greater than the standard, that period counts for only 1 day of exceedance, even if there may have been a second 8-hour exceedance in the same calendar day.

PM_{10} Correction for Missing Data When Evaluating the 24-Hour Standard Contrary to the gaseous standards evaluations, PM_{10} data are not normally required to be obtained on a daily basis. Unless the area is under special investigation or regulatory requirements, sampling once every 6 days is sufficient to evaluate the area for PM_{10} ambient air quality standard compliance. However, if less-than-every-day sampling occurs, the observed number of exceedances must be increased by the ratio of possible to monitored days in accordance with equation 4-2. This equation is used to evaluate the expected number of 24-hour standard exceedances.

The PM_{10} interpretive method is specified in 40 CFR, Appendix K, which indicates that annual PM_{10} compliance is evaluated on a quarterly basis to account for seasonal differences. In addition, the Appendix allows states to adjust data to account for exceptional trends and events. The equation for evaluating compliance with the 24-hour standard is as follows:

$$e_q = V_q\,(N_q/n_q) \tag{4-2}$$

where

e_q = expected number of exceedances in a calendar quarter, q
V_q = observed number of exceedances in the quarter
N_q = number of days in the quarter
n_q = number of days sampled with valid data in the quarter
q = the index for each calendar quarter (1, 2, 3, or 4)

To determine the expected number of 24-hour exceedances in a year, the 4 values of e_q are summed for the calendar year. The number of expected exceedances for each of 3 years is then summed and divided by 3 to determine the average annual expected exceedance number. If this number is less than or equal to 1, compliance is achieved.

It is obvious from equation 4-2 that, if only 1 out of every 6 days is sampled in a quarter (which is a standard monitoring frequency), and a single exceedance is observed in one quarter, the expected number of exceedances for that quarter will be 6. If the remaining three quarters of the year showed no exceedances, then the expected number for that year will also be 6 (i.e., $6 + 0 + 0 + 0 = 6$). If no exceedances are observed in each of the other 2 years, the average annual expected exceedance number over the 3-year period will be equal to 2 (i.e., $(6 + 0 + 0)/3 = 2$), and the site will be considered out of compliance, even though only a *single* value was *actually observed* to be over the standard.

In order to account for this type of situation and apparently unreasonable result, the regulations state that no adjustment for missing data will have to be applied if the state implements daily sampling immediately upon observing a single exceedance. Therefore, in the same example, if once the exceedance was observed, daily sampling was immediately implemented, and no further exceedances were observed in the year, the expected exceedance value for the year would only be 1, and the average for 3 years would be 0.33, thereby indicating compliance.

The implication of the above adjustment formula for missing PM_{10} data (equation 4-2) is that PM_{10} data should be obtained on as frequent a basis as possible, and daily is preferred.

Methods for Evaluating the Annual PM_{10} Standard To evaluate compliance status of the area with respect to the annual PM_{10} standard, four quarterly averages are calculated from the daily data, and these four quarterly means are then averaged to determine an annual average. Three annual averages are then averaged together and compared to the standard to determine compliance.

For example, assume that in one year there were quarterly averages of 52.4, 75.3, 82.1, and 63.2. The annual average for that year would be 68.3. If the second year had an annual average of 35.7 and the third year had an average of 42.3, the expected annual value would be equal to 48.8, rounded to 49. Since 49 is less than or equal to 50, the area is in attainment.

DESIGN VALUE CONCEPT FOR CLASSIFYING AREAS

In order to determine if an area meets the NAAQS for ozone and CO, the "highest second-highest" value must be evaluated. The term "high second-high value" is a shorthand method of referring to the fact that the NAAQS value cannot be exceeded more than once per year. However, the 1990 CAA also established a new concept, the design value, which is used to classify the severity of the nonattainment situation in an area. The design value for ozone is defined as the fourth-highest monitored value over the previous three-year period for which data are available. The design value for CO remains the same as the standard—that is, the highest second-high nonoverlapping 8-hour concentration in an area in one year.

The purpose of establishing ozone design values different from the standard is to remove some of the effects of unusual meteorological events on an area's classification. As is discussed in Chapter 9, various significant regulatory requirements are attached to the ozone nonattainment classification. Therefore, removing some of the meteorological variability is advisable.

ATMOSPHERIC DISPERSION MODELING

Compliance with the NAAQS and SAAQS for proposed sources or changes to existing sources of emissions is evaluated through ambient air quality dispersion modeling. The purpose of modeling is to determine what happens to pollutants when emitted into the atmosphere. When a pollutant is emitted, it mixes with the existing air and the concentration of the pollutant decreases as it mixes due to atmospheric turbulence. This process is termed *atmospheric dispersion*. The following paragraphs highlight some of the technical issues related to atmospheric dispersion modeling. There are many other issues not covered herein, and the reader should consult additional references, especially the *USEPA Modeling Guideline* and the model handbooks.

There are two keys to understanding the dispersion process:

1. The concentration of an emitted pollutant in the atmosphere must always decrease. The concentration can never be greater than when initially released.
2. The emitted plume will eventually reach the same momentum (i.e., speed and direction of motion) as the atmosphere into which it is emitted. Therefore, when a plume is emitted, the mixing process involves both momentum and constituent mixing.

Most of the point-source atmospheric models that are used in the regulatory process are termed *Gaussian dispersion models.* Although this sounds technically sophisticated, all that term means is that the distribution of pollutants in a plume can be described in terms of the Gaussian mathematical formulation. The Gaussian formulation is simply the standard bell curve used to describe normal distributions of any type of data. Remembering this fact allows one to understand how the models work and how various changes in model input parameters are likely to affect the results. Non-point-source models are used to assess regional pollutants (e.g., urban air shed modeling) or vehicle emissions (e.g., freeway or traffic impact modeling). These models are not discussed herein.

The following paragraphs discuss some of the technical issues related to atmospheric dispersion modeling.

Terminology

As a basis for discussing dispersion modeling, some of the commonly used terms are explained in this section.

Atmospheric Dispersion and Diffusion These terms are often used interchangeably. However, to physically understand the processes, it is useful to retain a distinction. The term "dispersion" refers to all of the processes that serve to dilute the plume as it is transported through the atmosphere by the local winds. The term "diffusion" refers only to the process of atmospheric mixing whereby the plume is mixed with ambient air due to atmospheric turbulence. Therefore, "dispersion" is a more encompassing term that includes "diffusion."

Plume Height The emitted pollutant plume will rise as a result of vertical momentum, both thermal (due to the plume being hotter than the surroundings) and mechanical (i.e., vertical velocity) momentum. Therefore, the height to which the plume rises is a function of the original physical height of the release point (e.g., the stack height) and the vertical momentum at release (i.e., temperature and velocity).

As the plume rises, it mixes with the ambient air that contains mostly only horizontal momentum. Therefore, as the plume mixes, the vertical momentum is dissipated and the plume bends over and travels horizontally with the ambient winds. The faster the horizontal winds are compared to the initial vertical velocity of the plume when emitted, the quicker the plume bends over horizontally and the lower the plume rise will be.

There are various formulae contained in the models that are used to calculate plume height. From a practical compliance standpoint, usually the only issue associated with plume rise is potential downwash related to a cavity wake effect for low stacks near buildings. Again, there are various formulae available in the models to calculate the downwash effect and incorporate that effect into the plume rise calculations. There are arguments over the accuracy of these downwash formulae, but the important thing to remember is that downwash is generally required to be examined in regulatory models.

Atmospheric Stability One of the factors that affects the plume rise in addition to the horizontal momentum is stability of the atmosphere. As a parcel of air rises, the air cools at the adiabatic lapse rate (about 1 degree Celsius per 100 meters for dry air) due to the decreased pressure (and increase in volume).

If the actual temperature of the atmosphere is greater at a given elevation than predicted by the adiabatic lapse rate, the atmosphere is stable and vertical motion is inhibited (because the rising air will be cooler than its surroundings). This is the condition characterized by atmospheric inversions. Conversely, if the actual temperature of the atmosphere is cooler than predicted by the adiabatic lapse rate, the atmosphere is unstable, and vertical motion occurs rapidly. If the temperature matches the lapse rate, the atmosphere is neutral, and vertical motion is neither enhanced nor inhibited.

There are different degrees of stability and instability, and the standard system of stability classification used in regulatory models allows for a total of six stability categories. Categories A, B, and C are for unstable conditions, with A being the most unstable. Category D is neutral, and categories E and F are stable, with F being the most stable.

Dispersion Coefficients For each stability category, dispersion coefficients can be assigned that indicate the amount of dispersion that will occur as the plume is transported in the atmosphere. Obviously, the more time (and, therefore, distance) that the plume travels, the more mixing will occur. Similarly, the more unstable the atmosphere, the more mixing will occur. Therefore, the dispersion coefficients are a function of stability class and distance (time).

Regulatory Defaults and Regulatory Models The standard atmospheric dispersion models that can be used to evaluate compliance with the CAA requirements are called *regulatory models*. These models are published in the EPA *Guidelines on Air Quality Models*, which is codified in 40 CFR Part 51, Appendix W. For each of the models, there is a certain set of model parameters and assumptions that must be used in order for the model to meet regulatory guidelines and requirements. In addition, local air quality agencies may establish certain parameter choices that must be used in the model. These choices are referred to as *regulatory defaults*. An example of a regulatory default is to incorporate downwash considerations into the model.

Source Types Sources of emissions are characterized as either point, area, or volume sources. Point sources are sources where the pollutant is emitted from a single stack, vent, or functionally equivalent opening. Area sources are those where the pollutant is emitted over a wide area; a landfill surface is an example of an area source. Volume sources are where the pollutant is emitted from a volume, such as a building; they represent conditions whereby the pollutants are initially emitted into the air within the building and then enter the ambient air through multiple vents, windows, and doors.

The reason for establishing different source types is that, when the receptor is close to the source, area sources will usually yield lower ambient concentrations than point sources and volume sources lower than area. This is to account for the fact that the chemical is already somewhat dispersed when emitted in the case of area and volume sources. As the receptor distance becomes greater (e.g., more than about 5 kilometers), the difference between a point, area, and volume source is negligible, because long-range atmospheric dispersion dominates over the initial dispersion caused by the area or volume source.

Principles of Atmospheric Dispersion Modeling

The fundamental equation that forms the basis of all the Gaussian dispersion models (Turner, 1970) is

$$C = \{Q/2\pi\sigma_y\sigma_z u\} \times \exp[-1/2(y/\sigma_y)^2] \times \{\exp[-1/2(z - h/\sigma_z)^2] + \exp[-1/2(z + h/\sigma_z)^2]\} \quad (4\text{-}3)$$

where

C = concentration (micrograms per cubic meter)

Q = emission rate of pollutant (micrograms per second)

u = horizontal wind speed at plume height (meters per second)

y = horizontal distance perpendicular to the plume axis (transverse direction, meters)

h = height of the plume centerline above ground (meters)
z = height of receptor above ground (meters)
σ_y = horizontal dispersion coefficient (meters)
σ_z = vertical dispersion coefficient (meters)

Figure 4-1 shows how the various parameters in equation 4-3 are related to plume dimensions and the Gaussian basis.

All of the parameters in equation 4-3 are physically obvious except for the dispersion coefficients. These coefficients, which are a measure of the dispersion of the plume, are a function of downwind distance. The most commonly used coefficients are called the Pasquill–Gifford (or P–G) coefficients and are found by consulting the graphs or equations shown in Appendix D.

The dispersion coefficients are empirically derived from field experiments and simply represent the standard deviation of the distribution of pol-

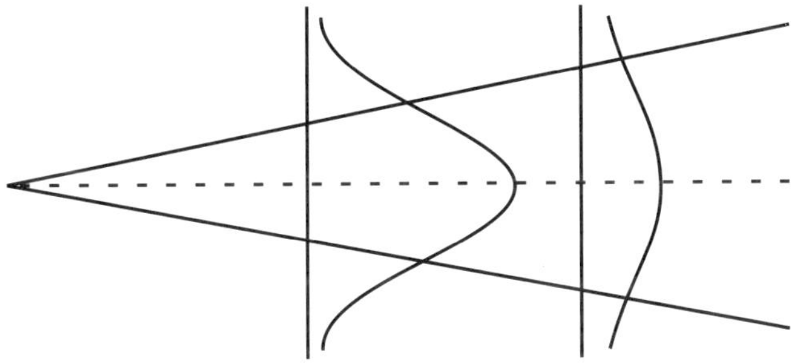

(a)

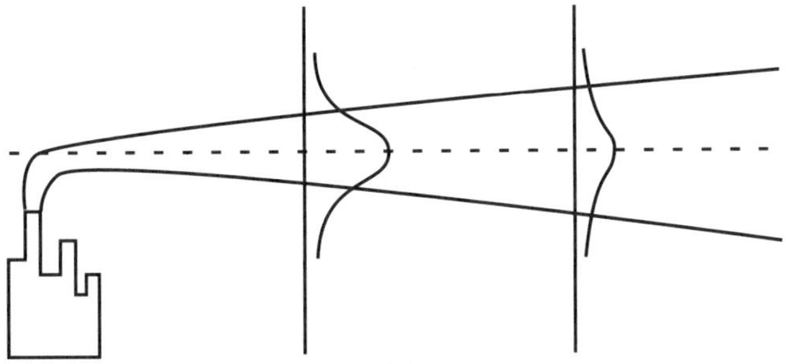

(b)

■ **FIGURE 4-1 Schematic diagram of Gaussian plume dispersion. (a) Plan view (horizontal cross section). (b) Vertical cross section. Note that horizontal dispersion is greater than vertical dispersion.** *Source:* After Turner (1970).

lutants in the plume. In a normal distribution (bell curve), 98 percent of the distribution is included when plus or minus 2.06 standard deviations are included (Turner, 1970). Therefore, the dimensions of a pollutant plume will be approximately four times σ_y and four times σ_z in the horizontal and vertical dimensions. It is useful to calculate these values when evaluating plume dispersion and assessing the validity of the results.

Selecting Stability Categories There are various schemes used to select the stability categories for a given meteorological situation. Table 4-4 is the most commonly used method that relies on cloud cover and wind speed observations.

Because cloud cover observations are not generally obtained in on-site meteorological monitoring programs, various other techniques for establishing stability class have been suggested and used. In addition, the graphical P–G dispersion coefficients shown in Appendix D have been reduced to equations that are a function of distance so that the values can be input into computer models. Therefore, commonly used models employ a combination of computer techniques to calculate the dispersion coefficients as a function of distance.

TABLE 4-4 Atmospheric Stability Classifications

A: Extremely unstable conditions
B: Moderately unstable conditions
C: Slightly unstable conditions
D: Neutral conditions[a]
E: Slightly stable conditions
F: Moderately stable conditions

Surface wind speed, m/sec	Daytime Insolation			Nighttime Conditions[b]	
	Strong	Moderate	Slight	Thin Overcast or >⅜ Low Cloud	< or = ⅜ Cloudiness
<2	A	A–B	B		
2–3	A–B	B	C	E	F
3–4	B	B–C	C	D	E
4–6	C	C–D	D	D	D
>6	C	D	D	D	D

[a]Applicable to heavy overcast day or night.
[b]The degree of cloudiness is defined as that fraction of the sky above the local apparent horizon that is covered by clouds.
Source: Hanna et al. (1982).

Wind Speeds Note that equation 4-3 needs wind speed *at plume height.* Wind speeds increase with height and can be calculated with a wind profile power law, shown in equation 4-4:

$$u = u_{10}(Z/10)^p \tag{4-4}$$

where

u = wind speed at plume height (meters per second)
z = plume height
u_{10} = wind speed at anemometer height (assumed 10 meters)
p = wind profile exponent

The values of p are shown in Table 4-5 for urban and rural conditions. The wind speed power law is built into the models. However, there is debate over the exact values of the power law exponents, and the models sometimes allow the user to select different values.

Calm Processing There are a few interesting consequences of the Gaussian formulation of equation 4-3. Note that if the release height of the plume is zero (i.e., a ground level release) and one wants to assess the centerline (i.e., maximum) concentration, equation 4-3 reduces to

$$C = Q/\pi\sigma_y\sigma_z u \tag{4-5}$$

This equation shows how the concentration in a plume is simply inversely related to wind speed and to plume dimensions. Therefore, if the

■ TABLE 4-5 **Wind Profile Power Law Exponents**

Pasquill Stability Category	Urban Wind-Profile Exponent *p*	Rural Wind-Profile Exponent *p*
A	0.15	0.07
B	0.15	0.07
C	0.20	0.10
D	0.25	0.15
E	0.30	0.35
F	0.30	0.55

Source: Turner, 1994.

wind speed is decreased by one-half, the concentration will double. The question arises: What if the wind speed is zero? The mathematics of equation 4-3 (or equation 4-5) will not allow the wind speed to be zero; and, in fact, atmospheric wind speed is not zero—there is always some horizontal air motion, even in extremely calm conditions.

To handle this situation, if the wind speed is reported as "calm" or reported as "zero," in the models the wind speed is assumed to be equal to 0.5 meters per second. This is called *calm processing*, and it is a setting in many regulatory models. If the calm processing option is not selected in a model, the model will not calculate a concentration for periods reported as zero or calm. Note also that the highest concentrations will occur when the wind speed is lowest. Therefore, ignoring calms in the models tends to yield lower concentrations.

Relative Concentration A common practice in modeling is to run the model with all parameters set to physical conditions, but enter a unitary emission rate (i.e., 1 gram per second). When this is done, the resulting concentration is called a *relative concentration* (C over Q). The modeled concentration of a given pollutant can then be calculated by multiplying the relative concentration times the actual emission rate. Because there is often more than one pollutant emitted from a single point, running the model once to obtain the relative concentration and then multiplying that value by the various pollutant emission rates saves a considerable amount of modeling time.

Complex Terrain Examination of equation 4-3 shows that the concentration of a pollutant in the plume increases as z approaches h (i.e., the receptor height approaches the centerline). The question arises: What happens when the terrain is above the height of the plume centerline? In this case, most of the models assume that the plume essentially impacts directly into the terrain, and the concentration is then double what it would be in the flat terrain case (i.e., when z is much less than h).

This is called the *complex terrain problem* ("complex terrain" being the term used to describe modeling when the terrain is above plume height). The EPA has now approved some additional modeling techniques, called *complex terrain modeling*, that allow the plume to travel around or over the terrain, depending on the relationship between atmospheric stability and wind speed. These models require more sophisticated data inputs, but avoid the concentration doubling problem that the simple models assume.

Complex terrain also greatly affects the local meteorology. Accounting for the effect of complex terrain on meteorological conditions, and, therefore, atmospheric dispersion, is somewhat involved and the source of much discussion between the regulatory agencies and the regulated community. Some of the adjustments include enhanced or diminished dispersion due to the physical conditions of a canyon and wind direction changes due to up-valley and down-valley flows (Hanna et al., 1982). The workbook by Hanna et al. contains an easily understood discussion of complex terrain effects on dispersion and attempts to quantify some of the effects.

Averaging Times Note that there is no averaging time explicitly included in equation 4-3. The averaging time for the concentration is built into the dispersion coefficients. When developed, the coefficients were representative of atmospheric diffusion over a 5- to 10-minute period (Slade, 1968). However, the regulatory models assume that the coefficients are representative of hourly diffusion (which overestimates the actual concentration). In the absence of complete meteorological data, the empirical conversion factors shown in Table 4-6 are applied to the modeled hourly concentration in order to arrive at concentrations for other averaging times. Note that the factors in the table account only for meteorological variability.

■ TABLE 4-6 Pollutant Averaging Time Adjustments

Modeling Technique	Averaging Time of Calculated Results	Desired Averaging Time	Multiplying Factor
Flat terrain—downwash	1 hour	3 hours	0.9
		8 hours	0.7
		24 hours	0.4
		Annual	0.1
Complex terrain	24 hours	1 hour	4.0
		3 hours	3.6
		8 hours	2.8
		24 hours	1.0
		Annual	0.4

Example: To convert a modeled 1-hour flat terrain concentration to a 24-hour average concentration, multiply by 0.4.
Source: CAPCOA (1989).

Another question regarding averaging time is, What averaging time is used for the emissions input value, *Q*? The regulatory agencies require one to use the maximum potential to emit (see Chapter 8) when assessing the regulatory applicability and impact of a facility. However, from an engineering perspective, the maximum emission rate is a function of averaging time. For example, the CO emissions of a boiler can fluctuate widely over a very short time (a few minutes) when the load is changed. On the other hand, over a 15-minute time period, the average emissions under full load will be about the same as those over a 1-hour time period (i.e., the CO emission rate is relatively stable).

Therefore, when inputting the emission rate, one should use an emission rate consistent with the model averaging time results intended. For example, if modeling to calculate an annual average concentration, the annual average emission rate should be used (either actual emissions or potential to emit, depending upon the application). If calculating an 8-hour average, an 8-hour emission rate should be used, and so on. Chapter 8 further discusses the emission rate averaging time issue in detail.

Deposition Modeling Equation 4-3 assumes that the pollutant, once emitted, remains airborne and remains unchanged over time. This is not the case for some pollutants. For example, particulates are deposited due to settling velocities. There are various mathematical schemes for handling such deposition; if this is occurring, such techniques should be used.

Regulatory Models As indicated, the EPA has published a Modeling Guideline that lists the various models that are approved for use in regulatory applications. The most recent edition of the Guideline was published in August 1995. The recommended models and their uses are shown in Table 4-7.

If one wants to use a model other than those in Table 4-7, it can be done, but advanced approval from the regulatory agency will be needed. The EPA Guideline lists a number of models that can and are routinely used in special applications (e.g., PLUVUE to calculate visibility impacts). However, when there is not a need for special modeling approaches, deviation from the regulatory models is usually not accepted by the EPA.

Commonly Used Gaseous Pollutant Models Of the list of models shown in Table 4-7, the most commonly used models are SCREEN (and the 1995 version SCREEN3) and ISC (Versions ISCST3 and ISCLT3, released in 1995).

TABLE 4-7 Summaries of Preferred Air Quality Models

Name	Abstract Abbreviation	Abstract Definition
Buoyant line and point source dispersion model	BLP	BLP is a Gaussian plume dispersion model designed to handle (a) unique modeling problems associated with aluminum reduction plants and (b) other industrial sources where plume rise and downwash effects from stationary line sources are important.
CALINE3—A versatile dispersion model for predicting air pollutant levels near highways and arterial streets	CALINE3	CALINE3 can be used to estimate the concentrations of nonreactive pollutants from highway traffic. This steady-state Gaussian model can be applied to determine air pollution concentrations at receptor locations downwind of "at-grade," "fill," "bridge," and "cut section" highways located in relatively uncomplicated terrain. The model is applicable for any wind direction, highway orientation, and receptor location. The model has adjustments for averaging time and surface roughness and can handle up to 20 links and 20 receptors. It also contains an algorithm for deposition and settling velocity so that particulate concentrations can be predicted.
Climatological dispersion model	CDM	CDM is a climatological steady-state Gaussian plume model for determining long-term (seasonal or annual) arithmetic average pollutant concentrations at any ground-level receptor in an urban area.

TABLE 4-7 Summaries of Preferred Air Quality Models *(Continued)*

Name	Abstract Abbreviation	Abstract Definition
Gaussian–Plume multiple–source air quality algorithm	RAM	RAM is a steady-state Gaussian plume model for estimating concentrations of relatively stable pollutants, for averaging times from an hour to a day, from point and area sources in a rural or urban setting. Level terrain is assumed. Calculations are performed for each hour.
Industrial–source complex model	ISC3	The ISC3 model is a steady-state Gaussian plume model which can be used to assess pollutant concentrations from a wide variety of sources associated with an industrial source complex. This model can account for the following: Settling and dry deposition of particles; downwash; area, line, and volume sources; plume rise as a function of downwind distance; separation of point sources; and limited terrain adjustment. It operates in both long-term (LT) and short-term (ST) modes.
Multiple point Gaussian dispersion algorithm with terrain adjustment	MPTER	MPTER is a multiple-point source algorithm. This algorithm is useful for estimating air quality concentrations of relatively nonreactive pollutants. Hourly estimates are made using the Gaussian steady-state model.
Single-source model	CRSTER	CRSTER is a steady-state, Gaussian dispersion model designed to calculate concentrations from point sources at a single location in either

■ **TABLE 4-7** *(Continued)*

Name	Abstract Abbreviation	Abstract Definition
		a rural or urban setting. Highest and high second-high concentrations are calculated at each receptor for 1-hour, 3-hour, 24-hour, and annual averaging times.
Urban airshed model	UAM	UAM is an urban scale, three-dimensional, grid-type numerical simulation model. The model incorporates a condensed photochemical kinetics mechanism for urban atmospheres. The UAM is designed for computing ozone (O_3) concentrations under short-term, episodic conditions lasting one or two days resulting from emission of oxides of nitrogen (NO_x), volatile organic compounds (VOC), and carbon monoxide (CO). The model treats urban VOC emissions as their carbon-bond surrogates.
Offshore and coastal dispersion model	OCD	OCD is a straight-line Gaussian model developed to determine the impact of offshore emissions from point, area, or line sources on the air quality of coastal regions. OCD incorporates overwater plume transport and dispersion as well as changes that occur as the plume crosses the shoreline. Hourly meteorological data are needed from both offshore and onshore locations. These include water surface temperature, overwater air temperature, mixing height, and relative humidity. Some of the key features include

TABLE 4-7 Summaries of Preferred Air Quality Models *(Continued)*

Name	Abstract Abbreviation	Abstract Definition
		platform building downwash, partial plume penetration into elevated inversions, direct use of turbulence intensities for plume dispersion, interaction with the overland internal boundary layer, and continuous shoreline fumigation.
Emissions and dispersion modeling system	EDMS	EDMS is a combined emissions/dispersion model for assessing pollution at civilian airports and military air bases. This model, which was jointly developed by the Federal Aviation Administration (FAA) and the United States Air Force, produces an emission inventory of all airport sources and calculates concentrations produced by these sources at specified receptors. The system stores emission factors for fixed sources such as fuel storage tanks and incinerators and also for mobile sources such as automobiles or aircraft. EDMS incorporates an emissions model to calculate an emission inventory for each airport source and a dispersion model, the graphical input microcomputer model (GIMM) to calculate pollutant concentrations produced by these sources at specified receptors. The GIMM, which processes point, area, and line sources, also incorporates a special meteorological preprocessor for processing up to one year of

■ **TABLE 4-7** *(Continued)*

Name	Abstract Abbreviation	Abstract Definition
		National Climatic Data Center (NCDC) hourly data. The model operates in both a screening and refined mode, accepting up to 170 sources and 10 receptors.
Complex terrain dispersion model plus algorithms for unstable situations	CTDMPLUS	CTDMPLUS is a refined point source Gaussian air quality model for use in all stability conditions for complex terrain applications. It contains, in its entirety, the technology of CTDM for stable and neutral conditions. However, CTDMPLUS can also simulate daytime, unstable conditions and has a number of additional capabilities for improved user friendliness. Its use of meteorological data and terrain information is different from other EPA models; considerable detail for both types of input data is required and is supplied by preprocessor and the association of each model receptor with a particular hill.

The SCREEN models are single-source models (i.e., only one emitting type can be modeled at one time). (In 1995, the EPA issued the SCREEN3 model, which is a slight modification of SCREEN. However, the models operate in essentially the same way and yield the same results. Therefore, in this discussion, the model will be referred to only as "SCREEN.") The advantage of SCREEN is that it does not require any meteorological data input. SCREEN works by inputting the theoretical worst-case meteorological conditions that could ever occur, regardless of whether or not they do occur, and then calculates the highest concentration during those worst-case conditions. Therefore, SCREEN yields a conservative upper-bound estimate of the ambient concentrations that could occur from a facility.

The problem with SCREEN, when it is applied in a regulatory setting, is that the regulatory agency and industry tend to forget the fundamental premise of SCREEN (i.e., it predicts the theoretically highest concentration that could ever occur). When this fact is overlooked, the SCREEN results are incorrectly interpreted as representing ambient impacts and overly stringent regulatory decisions (or overly stringent management decisions) are made.

ISCST3 and ISCLT3 use the same principles as SCREEN, but they input actual meteorology. The ISC model handles multiple sources in any combination. Therefore, ISC is more flexible. [For convenience, ISCST and ISCLT will be collectively referred to as "ISC." The ST version yields short-term as well as long-term (i.e., annual average) concentrations. ISCLT yields only long-term averages.]

In the past, the ISC model was difficult to run on personal computers because of the large amount of computations required. Therefore, many practitioners used SCREEN to the exclusion of ISC. However, with modern personal computer processing speeds, if meteorological data are available, ISC is almost as easy to run as SCREEN, and ISC yields much more representative values. Therefore, although SCREEN is the most commonly used model, ISC should not be ignored in most cases.

Particulate Models Particulate impacts are commonly modeled with the SCREEN or ISC model, just as for gaseous pollutants. This common practice tends to greatly overestimate the ambient concentrations of particulates, especially in the case where total suspended particulate (TSP) ambient concentrations are being assessed. This is the case because the ISC and SCREEN models do not account for particle deposition, which tends to lower the ambient air concentrations (deposited particles are not breathed into the lungs). The best regulatory model for particulate impact assessments is the FDM model. This model requires a significant amount of input data, especially with respect to particle size distributions. However, the model does contain defaults, and the relatively greater accuracy of the model is an acceptable price to pay in many particulate impact evaluations.

Accuracy and Precision of Models Under the best of circumstances that precisely match the assumptions built into the Gaussian formulation (i.e., flat terrain, constant wind conditions, 5- to 10-minute averaging times), the models predict concentrations within about a factor of two of actual observations. However, the regulatory models and parameter assumptions and defaults used often cause the models to overpredict concentrations by factors of 10 or more. Therefore, it is not unusual for monitoring results to be much, much different than modeled results. However, EPA and local agencies routinely use models on which to base their permitting and other air quality decisions.

Transport Time The EPA requires impact analyses to be conducted at least to 50 kilometers from the source, and often 100 kilometers away. One fact that is often overlooked is that the amount of time that it takes to transport a plume is significant. For example, if the wind speed is 2 meters per second (a common wind speed for high concentration situations), the transport time to 100 kilometers will be 50,000 seconds (i.e., 100,000 meters divided by 2 meters per second); 50,000 seconds is nearly 14 hours. A 2-meter-per-second wind in a straight line is not likely to occur for 14 hours. Obviously, when evaluating model results, transport time must be considered.

Pollutant Transformations Some pollutants undergo chemical transformation in the atmosphere. The most commonly considered transformation is NO_x to NO_2 (Chapter 4). The most recent USEPA modeling Guideline recommends a tiered screening approach. Under this approach, it is initially assumed that all NO_x emitted is completely and instantaneously converted to NO_2. This provides a conservatively high upper-bound estimate of NO_2 concentrations. If this first-tier assumption indicates possible exceedances, the Ambient Ratio Method (ARM) is used to further refine the estimated concentration. In the ARM method, the tier 1 results are multiplied by 0.75 to calculate NO_2 concentration. The 0.75 factor is a national average of observed NO_2 to NO_x ratio. If the ARM approach is still too conservative, a case-by-case modeling analysis using site-specific data and a reactive plume model can be used to estimate concentrations of NO_2. This tiered approach is new (in 1995). The use of a nationwide ARM value is also new, but it should yield results similar to previously recommended methods (i.e., the Ozone Limiting Method).

Background Concentrations When modeling, a common mistake is to forget to add the existing background concentrations to the calculated ambient concentration from the model. In order to assess compliance with the NAAQS, the modeled concentrations must be added to the background, because it is the total pollutant concentration that potentially causes health effects. However, when evaluating PSD increments, by definition (see Chapter 8), background concentrations are *not* added to the modeled results.

A second common mistake involving background concentrations is to add the maximum modeled concentration at a certain receptor (e.g., the maximum theoretical concentration calculated by SCREEN) to the maximum measured background concentration that was measured at a different location. Although this procedure does yield an upper bound maximum value (i.e., in no case can the actual combination of measured and modeled values be greater), it is scientifically incorrect and the upper bound value calculated by this procedure often grossly overestimates the ambient air quality impact of the source. When adding background concentrations, only those modeled and measured values at the same location can be summed together.

SUMMARY

Air quality impact modeling is complex, but it is understandable. The key is to recognize that the computer models are based on physical principles. Therefore, the model must be adjusted so that the modeled situation matches the actual situation as closely as possible. One must not just run the model with standard defaults. One must review the model inputs and outputs and attempt to ensure the representativeness of the models.

CHAPTER 5

RISK ASSESSMENT AND MANAGEMENT

The previous chapter discussed evaluation of ambient air quality impacts of sources with respect to national and state ambient air quality standards. These standards are specific values established by the regulatory agencies in order to protect the public health and welfare with an adequate margin of safety. These types of standards are termed "bright-line" standards. Bright-line standards mean that above the line (i.e., the standard) the public is not safe whereas below the line the public is safe.

However, for a very large number of air pollutants—for example, the air toxics or hazardous air pollutants (HAP)—there is no bright line. This is because it is assumed that for cancer-causing chemicals (i.e., carcinogens), there is no safe level and "any amount" of a carcinogen could cause cancer. In order to assess and regulate sources of HAPs, the agencies use risk assessment and risk management.

DIFFERENCES BETWEEN RISK ASSESSMENT AND RISK MANAGEMENT

Risk assessment is the process of evaluating and estimating the risk to the public from HAP releases. Guidelines and methods to perform risk assessments are suggested by various agencies and result in numerical estimates of the risk of HAP exposure.

Risk management is the decision-making process based on the risk assessment results. The risk assessment process contains numerous, very conservative assumptions that cannot be verified as being valid. Therefore, in the risk management process the regulatory agency must evaluate the risk assessment results and consider other factors (e.g., actual population exposure) to make decisions related to the specific source of HAP (e.g., to require additional controls).

In practice, risk assessment is often combined and confused with risk management. The regulatory agencies generally rely on the risk assessment results without much consideration of other factors. This is generally overly conservative and usually results in more stringent decisions than necessary. The agencies do this, however, because of the lack of resources and legislative authority to make alternative risk management decisions. This is especially the case at the local level.

RISK ASSESSMENT COMPONENTS

Risk assessments consist of four distinct components:

1. Hazard identification
2. Dose-response assessment
3. Exposure assessment
4. Risk characterization

The hazard identification step is the process of determining the potential health effects that may be associated with emitted HAPs. The step requires identification of individual HAPs and the emission rates. The time variability of emissions rates (e.g., batch processes whereby relatively large emissions are released over a short time versus continuous processes where emission rates are relatively constant) must also be determined. When the HAPs are identified, the potential health effect (e.g., cancer versus noncancer effects) for each HAP is also identified.

The dose–response step establishes the dose–response values that will be used in the risk assessment. For cancer effects, the dose–response relationship is a derived potency relationship between dose and effects that is used to calculate the probability or risk of contracting cancer when exposed to a given concentration of a HAP. These potency relationships are derived from animal research data. For noncancer effects, dose–response data developed from human or animal studies are used to establish acceptable exposure levels (AELs). The AELs are used to calculate a hazard index (HI) which is the ratio of the modeled (or actual) HAP concentration to the AEL.

The exposure assessment step is used to estimate the extent of public exposure to each HAP. This involves air quality dispersion modeling and exposure route analyses (e.g., inhalation, ingestion of contaminated soils, mother's milk, or dermal absorption).

Risk characterization is the process of characterizing the magnitude and extent of the risk. In this step, the nature and amount of exposed population and multiple effects from multiple HAPs (including background concentrations if applicable) are assessed. This final step integrates the health effects assessment and public exposure information and forms the basis for risk management decisions.

RISK ASSESSMENT TECHNIQUES

Hazard Identification

The hazard identification step of the risk assessment (RA) process is to identify the type and quantity of emitted chemicals and the potential health effects associated with the chemicals.

Estimating Emissions Estimating emissions is always a difficult process, and for criteria pollutants the emission estimating process is verified through source emissions testing (sometimes called "stack testing"). However, very often, hazardous air pollutant emissions are not amenable to source testing, either because there are no published methods for the testing or the source of the emissions cannot be tested (e.g., when the source is a fugitive emission). Therefore, even this relatively simple step of estimating emissions is fraught with uncertainty.

In the absence of source emission testing, emission factors are used. These factors calculate the mass of emissions given a certain process rate and other physical parameters associated with the emitting device or process. Many agencies and the EPA have published emission factors for air toxics.

The toxic release inventory (TRI) is at times used to derive emission factors for toxics. The TRI is compiled as required by the Superfund Amendments and Reauthorization Act of 1987 (SARA) Title III. These emission reports, reported to EPA on Form R, provide estimates of the annual amount of chemicals released to the air, water, and soil by industries that manufacture or use greater than threshold amounts of chemicals.

The problem with using the TRI, however, is that the reported releases are usually overestimates. This is because, for purposes of the TRI, there is no negative impact to the reporting facility for overestimating releases, although there *is* a negative impact if the industry incorrectly underestimated the releases. Therefore, use of the TRI to derive toxic emission factors is suspect.

Another technique routinely used to estimate toxic emissions is mass balance. If one knows with certainty all of the inputs and outputs of the process stream, mass balance is a precise method of estimating emissions. However, in general, the facility does not know all inputs and outputs. Usually, only the input function is known with precision (i.e., how much chemical was purchased and entered into the processing system in a given time period). For example, many chemicals undergo chemical reactions in the manufacturing process or may be absorbed into plant equipment and structures. Therefore, just as in the case of the TRI data, mass balance is often not very precise and tends to overestimate emissions. If mass balance is used, careful consideration of all possible outputs must be made. Although mass balance is relied on extensively in the air toxics arena, emission factor and source testing techniques are preferred.

Air Toxics Source Testing If source emissions testing is to be used to determine toxic emissions, detection limit problems must be carefully analyzed. For many chemicals, the toxic health effect of interest occurs at very low concentrations. Therefore, if the detection limit is not sufficient, even though a chemical may not be detected in the stack emissions, an assumption is often made that the chemical could have been present at some level less than the detection limit.

The common assumption is to assume that the chemical was emitted at the detection limit value or at one-half the detection limit value. Such assumptions can cause inappropriate risk assessment results and risk management decisions. Therefore, when performing a source test for toxics, the required minimum detection limit should be back-calculated (using the techniques described later in this chapter under Risk Screening Techniques). When the necessary detection limit has been determined, the source testing methods can be adjusted as necessary to meet the limit (usually by increasing the duration of the test).

The second problem encountered in source testing for toxics is that of testing methods. The EPA and various local regulatory agencies publish source testing methods that must be used for compliance testing. Most of these methods are for the criteria pollutants, not air toxics. Therefore, one must adapt a method and/or develop a new method for the specific toxic and device of interest.

Testing methods published by the National Institute of Occupational Safety and Health (NIOSH) are often used as surrogates for source emissions testing methods. But this common practice is very uncertain and should be avoided. The NIOSH methods are published for obtaining workplace exposure data for chemicals usually observed in considerably higher concentrations than will be found in a stack. In addition, workplace exposures are usually at room temperature, and the air in which the chemical of interest is mixed is essentially clean air. In contrast, stack emissions are usually at high temperature and contain many other chemicals at relatively high concentrations. Therefore, there can be considerable chemical interferences and other problems causing the NIOSH methods to fail when used for source testing. If a published stack emission testing method is not available, prior to using a NIOSH method, it should be evaluated by an experienced environmental chemist.

A third problem is possible atmospheric chemical transformations. Potentially toxic chemicals are often relatively reactive in the atmosphere or in the exhaust stack. Therefore, when conducting the later stages of the RA process, one may overestimate the concentrations of the emitted chemical or underestimate concentrations of reaction products if chemical transformations are not taken into account. Likewise, it is possible that the emitted chemical transforms into a more toxic chemical in the atmosphere. This possibility should also be examined in the hazard identification process.

Types of Health Effects Health effects can be generally categorized as direct and indirect. The direct health effects are those that are directly observable—for example, coughing and watering eyes that occur when exposed to high levels of ammonia. Other health effects, such as effects on the central nervous system, blood, gastrointestinal tract, respiratory tract, kidney, liver, skin, immune system, and reproductive system (including developmental abnormalities) are also included in this category.

Indirect health effects, on the other hand, are those that are not directly observable. The main example of this type of effect is carcinogenesis.

Health Effects Databases After the type and quantity of emissions are determined, potential health effects of the emitted chemicals must be identified. There are several reference databases that publish lists of chemicals and possible health effects at various doses. The key to successfully using these databases is to remember the following:

1. Chemical toxicity is dose-dependent (i.e., too much of anything is toxic, even water).
2. The scientific basis for listing the chemical and health effect must be thoroughly understood.

One of the common references is a booklet published by the American Council of Governmental Hygienists (ACGIH) entitled *Threshold Limit Values and Biological Exposure Indices for 1992–1993* (ACGIH, 1993). However, this booklet lists worker-based health effects levels and information, not ambient exposure health effects. In fact, the ACGIH itself states that the information in the booklet cannot be used to establish ambient air quality standards. The reason it is used is that it is widely available, is easily reviewed (it is only about 125 pages long), and lists over 800 chemicals.

Probably the most thorough database to use is the Integrated Risk Information System (IRIS) published by the EPA. IRIS is accessible by computer as well as in hard copy. The advantage of IRIS is that, before a chemical and health effect is listed, the Agency reviews the scientific literature related to that chemical and concludes that the health effects noted are possible. However, even in IRIS, the quality of the scientific literature used as the basis for listing is subject to variability. Therefore, even in IRIS, one should critically review the information prior to accepting the health effects at the levels noted, and the services of an experienced toxicologist should be sought to verify or refine the published values.

Regulatory Health Effects Data Practically speaking, when assessing the impact of a facility, the regulatory agency has already identified the potential health effects related to a given chemical. Therefore, normally, the most onerous effort in this stage of the RA is determining the emission rates.

Duration of Health Effects Health effects occur when there is sufficient intensity (concentration or dose) and duration. In general, duration is categorized as acute or chronic. Chronic duration is long-term, generally evaluated as exposures over a yearly period. Acute exposure is short-term, generally evaluated as exposures over a one-hour period. There are subchronic and subacute exposures as well, but for purposes of ambient air quality risk assessments, only the hourly and annual exposure durations are normally evaluated.

Dose–Response

The second step of the risk assessment process assigns a dose–response value to each of the emitted chemicals. These values are in terms of "unit risk" for cancer effects and acceptable exposure levels (AELs) for noncancer effects.

Dose–response values are developed from observations and extrapolations of laboratory animal tests and/or human epidemiological data. In the case of cancer, the extrapolations occur through a zero threshold linear model, called the *one-hit model.* The one-hit model means that if one molecule of the chemical interacts with a single cell in the body (i.e., "one hit"), there is a greater-than-zero probability that the interaction will cause cancer. This means that only if the chemical exposure is exactly zero is there a zero chance of cancer. This is why the model is called a *zero threshold model.* The model is also a linear model, which means that if the dose of chemical is doubled, the probability of contracting cancer is also doubled.

For the noncancer effects, extrapolation of laboratory animal tests is conducted through various similarity models that relate the biological processes of the laboratory animal to the human. When extrapolating these data, one must be careful to ensure that the health effect observed in the laboratory animal is capable of occurring in humans. For example, rats will get liver damage from certain chemicals because of a damage to a unique enzyme that exists only in rat livers. Therefore, this chemical will not cause liver damage in humans.

The services of an environmental toxicologist are often necessary in order to review potential health effects data and establish dose–response levels. Environmental toxicologists understand the difference between the extremely high exposures that laboratory animals are subjected to and the exposures likely to occur in the environment.

For carcinogenic chemicals, dose–response is expressed in terms of the probability of contracting cancer as a result of exposure to a given chemical. This is an incremental risk—that is, the specific risk from the given chemical in addition to the risks already endured by the population. The probability of contracting cancer is expressed in terms of additional cancers per numbers of exposed persons. For example, if exposure to a chemical would cause 2 additional cancers in a population of 1,000,000 exposed persons, the incremental risk would be stated as 2 in one million or 2×10^{-6}. In a population of 1,000,000, about 300,000 contract cancer. Therefore, an incremental risk of 2 in 1,000,000 means that there would be 300,002 cancers out of a million (i.e., 2 additional).

As indicated, for many chemicals, the regulatory agencies have evaluated the health data and already published dose–response values. One of the more complete such compilations is published by the California Air Pollution Control Officers Association (CAPCOA, 1992). These values have been reproduced in Appendix E for the federally listed hazardous air pollutants (see Chapter 7).

Appendix E provides unit risk values for carcinogenic chemicals and AELs for those chemicals that have noncancer effects. (Note that some chemicals have both types of effects.) Unit risk values are the probability of contracting cancer if one is exposed to 1 microgram per cubic meter (a "unit" exposure) of the chemical for a 70-year lifetime.

The AELs are the level at which direct health effects could occur in the most sensitive populations. For noncarcinogenic chemicals, the exposure is assessed as a ratio of exposure to the AEL. This ratio is termed the *hazard index* (HI). If the HI is less than 1, there would not be an expected health effect. If it were greater than 1, a health effect could possibly occur.

Exposure Assessments

Determining Dose for Single-Pathway Chemicals In order to determine the risk, the dose–response values must be coupled with estimates of chemical exposure. Air quality dispersion models are used to estimate the ambient concentrations of the emitted chemicals. Hourly average and annual average concentrations are calculated in order to evaluate both acute and chronic health effects. After the concentrations are calculated, they must be converted into dose.

For single-pathway carcinogenic pollutants, whose only route of exposure with potential health effects is inhalation, the dose is simply the concentration of the chemical in the ambient air times the amount of air breathed. It is generally assumed that residential persons breathe 20 cubic meters per day. For example, if the concentration of a chemical is 5 micrograms per cubic meter, the dose due to inhalation would be 100 micrograms per day (i.e., 5 micrograms per cubic meter times 20 cubic meters per day). To calculate a lifetime dose, it is assumed that a person breathes the chemical continuously for 70 years.

Note that the unit risk factors in Appendix E are in terms of concentration (i.e., micrograms per cubic meter), not dose (i.e., micrograms per lifetime or micrograms per day). This is because the agency publishing Appendix E has already applied the "20 cubic meters of air breathed per day for 70 years" assumption to the potency values in order to arrive at the unit risk factors shown in the Appendix.

Determining Dose for Multipathway Chemicals For multipathway carcinogenic pollutants, calculation of dose is considerably more complicated. Figure 5-1 shows the various routes of exposure. For each of these routes there is a set of equations that calculates the dose through that route.

For example, metals such as hexavalent chromium (hex chrome) emitted into the air can be deposited onto the soil. In order to calculate the dose through this exposure route, an assumed amount of soil ingested per day would have to be multiplied by the concentration of hex chrome deposited on the soil. The amount deposited is a function of the concentration of hex chrome in the air and the deposition velocity of the hex chrome. The deposition velocity is a function of the particle size and the surface roughness of the soil. These various parameters either are established by the regulatory agency or must be derived by the risk assessment practitioners.

Adjusting Dose for Exposure Conditions Depending on the person exposed, the dose calculation needs to be adjusted. The standard dose assumption is that a person is exposed continuously for a 70-year lifetime. However, if the exposed person is a worker at a commercial establishment, it is obvious that the person does not work continuously for 70 years. In this case, the agency allows the exposure to be adjusted by the number of hours worked in a lifetime compared to the number of hours in a 70-year lifetime. This adjustment, termed the *worker adjustment factor*, can be derived as follows.

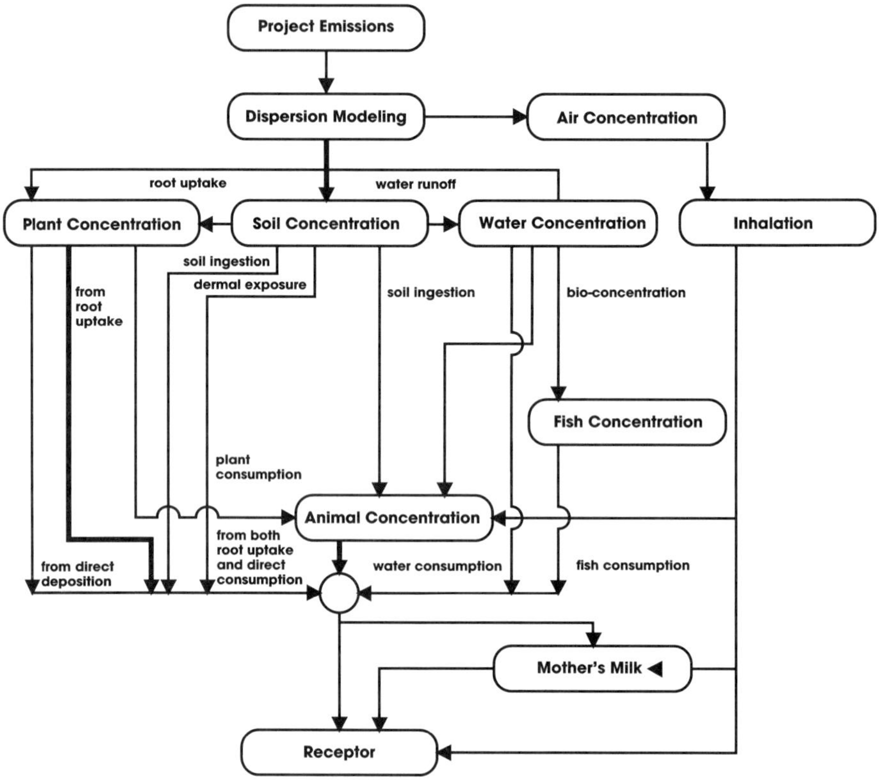

■ **FIGURE 5-1 Multipathway exposure routes.** *Source:* CAPCOA (1989).

For a worker, the California Environmental Protection Agency (California Code of Regulations, Section 12721, Proposition 65) assumes that a worker works 40 years, 5 days per week, 50 weeks per year (allowing for 2 weeks' vacation). However, it is also assumed that the worker breathes half of his or her daily intake of air during the days they work (one breathes more air while working than while resting). Therefore, from a dose basis, the worker exposure should be considered equal to 12 hours of residential exposure on the working days.

This assumption provides for 120,000 hours of exposure in a lifetime (12 hours per day times 5 days per week times 50 weeks per year times 40 years). For a nonworker, as stated, it is assumed that the person is exposed for 70 years, 365 days per year, 24 hours per day, for a total of 613,200 hours (24 hours per day times 365 days per year times 70 years). The ratio of worker exposure hours to total hours is 120,000 to 613,200, or 0.196, thus the adjustment factor.

This agency also allows an adjustment factor for the casual visitor. In this case, it is assumed that the visitor visits the facility 1 hour per month, 12 months per year, for 70 years. It is also assumed that the visitor breathes at the same rate while visiting as while in his or her residence. This results in a ratio of visitor hours to lifetime hours of 0.0014.

In order to assess the risk for workers versus residents, the residential risk value calculated from the 70-year dose (i.e., unit risk factor) is multiplied by the adjustment factor of 0.196. Likewise for visitors, the risk is equal to the 70-year dose times 0.0014.

Linear Nature of Dose Calculations As the above paragraphs regarding exposure duration adjustment factors demonstrate, the risk models and dose–response relationships for carcinogenic chemicals are linear. Therefore, if the exposure duration doubles, the risk doubles; if the concentration of a chemical is one-third as great as previously, the risk is one-third as great. This is a key point to understanding and working with risk assessments: Once the dose–response model is established, the cancer risk is linear with respect to chemical concentrations.

There is one caveat with respect to the linearity assumption. Some of the multipath exposure equations have nonlinear components. However, even in these cases, once the initial dose–response values are input and the risk is calculated, the resulting risk value is linear with respect to concentrations.

Dose for Noncancer Health Effects The AELs for noncarcinogenic chemicals are in terms of concentrations, not dose. This is because, for direct health effects, the effect is not linear. In other words, if the concentration of the chemical is less than the AEL, it is assumed that there will be no health effect; if the concentration is greater, it is assumed that the health effect will occur. Even if the concentration is slightly or greatly greater than the AEL, it is still assumed that the same health effect will occur.

Risk Characterization

When the risk of individual exposed persons (i.e., workers, residents, visitors) is calculated, the risk characterization step is completed. The key terms in this step are

1. Maximum exposed individual resident (MEIR)
2. Maximum exposed individual worker (MEIW)
3. Actual exposed person (AEP)
4. Cancer burden

The MEIR is the maximum risk at any point near the source, assuming that a person lives there for 70 years under continuous exposure to the highest modeled ambient concentration of the pollutant. The risk at this point is assessed and often used as a basis for risk management decisions even if no one actually lives at that point (e.g., the point is an empty lot).

Likewise, the MEIW is the maximum risk at any point where workers are exposed. It is important to note that the workers assessed here are off-site workers, not workers at the facility. This is because of the definition of ambient air. For the MEIW, the worker adjustment factor is applied as discussed previously.

On the other hand, some agencies are evaluating risk and making risk management decisions based on the AEP. The AEP is an assumption based upon the exposure to actual persons living in the area of the facility. For example, rather than assuming that a person lives for 70 years in the same location, it could be assumed that the person lives there for only 7 years, and thus the risk would be one-tenth as great. Another change is to calculate risks only where there is or could be residences (e.g., risk for receptors on highways would not be calculated). Although the AEP concept is intellectually appealing, it is not generally applied until the risk management decision stage (i.e., the regulatory agencies usually require MEIR and MEIW assessments and reports).

The cancer burden is the number of additional cancers that could occur in the area affected by a facility. Cancer burden is calculated by multiplying the risk times the number of exposed persons. For example, if 50,000 persons were exposed to a risk of 5×10^{-6} (i.e., 5 in 1,000,000), the cancer burden would be 0.25 (i.e., $5 \times 10^{-6} \times 50{,}000$).

RISK MANAGEMENT

Although the risk assessment process is built on the principle that there is no safe level of pollutant for carcinogenic chemicals, in practice, a "safe" level is often *de facto* established through the risk management decisions made by agencies.

The range of acceptable cancer risks is from 1×10^{-4} (100 in 1,000,000) to 1×10^{-7} (0.1 in 1,000,000) per EPA guidance. However, most agencies use 1×10^{-6} (1 in 1,000,000) as a point of departure for decision making. Generally, if the risk is less than 1 in 1,000,000, no further action to reduce the risk is required. If the risk is 10 in 1,000,000, some sort of risk reduction process is required to attempt to get the risk below 1 in 1,000,000. If the risk is greater than 100 in 1,000,000, the agencies generally will not permit the source or will require action to immediately reduce the risk. Therefore, in practice, there is often a *de facto* bright-line standard, usually 1 in 1,000,000.

Acceptable cancer burdens are usually 0.5 or less in the exposed population, when the exposed population is considered to consist of all persons exposed to an MEIR or MEIW greater than 0.1 in 1,000,000.

With respect to the noncarcinogenic chemicals, a hazard index (HI) of less than 1 is generally considered acceptable. However, the agencies often view an HI greater than 0.5 (i.e., one half the AEL) to be cause for additional action or investigation.

RISK SCREENING TECHNIQUES

A formal risk assessment requires a considerable expenditure of resources. Therefore, it is often beneficial to employ a risk screening technique prior to conducting a formal risk assessment. The purpose of risk screening is to identify those emitted chemicals that pose the greatest risk. After these chemicals are identified, early risk management decisions (e.g., chemical substitution, additional controls) can be implemented. As will be seen in Chapter 7, risk assessments are used by the EPA to develop additional emission controls. Therefore, the ability to perform risk screening is important.

There are two levels of risk screening, depending on the amount of site-specific data input into the screening methodology:

1. Site-specific emissions data only, or
2. Site-specific emissions data and emissions characteristics (i.e., data necessary to input into the SCREEN model)

In order to carry out the first type of screening, generic emissions characteristics are assumed so that the only input data needed are the mass emissions rate and the dose–response value (either unit cancer risk or AEL) of each chemical. This type of method has been published in California as part of the California Air Toxic Hot Spots program (CAPCOA, 1990).

The basis for this screening procedure is the following formula:

$$RS = Q \times URV \times 1700 \tag{5-1}$$

where

RS = risk score

Q = emissions rate in pounds per year

URV = unit risk value (Appendix E)

1700 = conversion and dispersion factor

In this equation, the resultant RS is equal to the risk per 1,000,000. In other words, if RS equals 10, the risk is equivalent to 10 in 1,000,000.

For noncarcinogenic chemicals, the screening equation is

$$RS = (Q/AEL) \times 150 \text{ for chronic effects} \tag{5-2}$$

or

$$RS = (Q/AEL) \times 1500 \text{ for acute effects} \tag{5-3}$$

where

RS = risk score

Q = emissions in pounds per year for chronic effects and pounds per hour for acute effects

AEL = the acceptable exposure level (Appendix E)

150 and 1500 = conversion and dispersion factor

In this equation, if the RS is greater than 10, it indicates that the AEL is likely to be exceeded, and if less than 10, the AEL is not likely to be exceeded.

The above equations are very conservative and overestimate the possible risk. If one wants to refine the screening estimates, the next step is to obtain emissions characteristics data and run the SCREEN model. Once the SCREEN concentrations are obtained, they can be compared directly to the AELs for determining a hazard index (HI) for noncarcinogenic chemicals. For carcinogenic chemicals, the SCREEN concentrations (in micrograms per cubic meter) can be multiplied times the unit risk values of Appendix E in order to calculate a risk. In this case, if the HI is greater than 1 or the calculated cancer risk is greater than 1 in 1,000,000, it is possible that further risk reduction efforts or a refined risk assessment may be required.

CONSERVATISM IN RISK ASSESSMENT TECHNIQUES

The risk assessment techniques that are used in the regulatory process contain numerous conservative assumptions. These assumptions are multiplicative and result in overestimates of risk by factors of at least 1000. For example, the SCREEN dispersion model overestimates ambient concentrations by about a factor of 10. The exposure assumptions are at least 10 times too great (70 years in one place versus 7 years) and the unit risk values derived from the one-hit model overestimate risk by at least another factor of 10. Each of these steps multiplies in the risk assessment process and yields at least 1000 times too high of a risk estimate. One of the results of the conservative methods, however, is that risk assessments are generally conducted *without* adding background concentrations.

CHAPTER

6

EMISSION STANDARDS FOR CRITERIA POLLUTANTS

NEW SOURCE PERFORMANCE STANDARDS (NSPS)

The CAA requires EPA to promulgate emission limits for new and modified sources of emissions. These emission limits are termed New Source Performance Standards (NSPS). Although these emission limits are termed performance "standards" by the Act, it is good practice to avoid using the term "standard" to refer to an emission limit. The best practice is to use the term "standard" to refer to ambient air concentrations (i.e., the NAAQS) and the term "emission limit" to refer to device-specific emission limits (i.e., the NSPS).

The NSPS are promulgated by the EPA on a device- and pollutant-specific basis and apply to all new and modified sources for which an NSPS has been promulgated throughout the nation. (Note that existing sources are regulated through other state-specific programs established by the SIPs.) The NSPS are promulgated for specific source categories that EPA judges to cause or contribute significantly to air pollution that may be reasonably anticipated to endanger public health or welfare [Section 111(b)(1)(A) of the Act].

The NSPS must reflect the degree of emissions limitation and the percentage reduction achievable through the application of the best technological system of continuous emission reduction that EPA determines are adequately demonstrated for each category of sources. This determination must take into account the cost of achieving such reductions and any non-air-quality health, environmental impact, and energy requirements (Section 111(a)(1)).

Note that intermittent control systems cannot be used as an NSPS, because the law requires a continuous emission reduction system. The NSPS must involve a technological system and may *not* include nontechnical solutions such as fuel switching or work practices. (This is in contrast to the MACT standards for air toxics, which can include work practices. See Chapter 7.)

In addition to specifying emission limits, the NSPS can, and often do, include provisions for monitoring and other recordkeeping to demonstrate compliance with the emission limits. The NSPS also often require compliance tests to demonstrate that the new or modified equipment meets the NSPS. These tests are often required within 180 days after startup of the equipment. It is important to note that the NSPS compliance test requirements and procedures are in addition to any requirements the local permitting authority may place on the facility. There have been occasions when the state approved an emissions test as meeting the permit conditions, but because the methodology was not consistent with the NSPS, EPA issued an enforcement action against the facility for not conducting a compliance test as required by the NSPS.

The list of source categories for which NSPS have been promulgated as of August 1, 1995, is included in Appendix F. Three examples of NSPS limits follow.

Example NSPS—New Electric Generating Units

For "new" utility steam generating units, the NSPS are specified in 40 CFR 60 Subpart Da. New units are those constructed after September 18, 1978. (40 CFR 60 Subpart D established less stringent NSPS limits for units constructed before September 18, 1978.) The new units for which Subpart Da applies are called "Subpart Da" plants. However, Subpart Da applies only to electric steam generating units capable of combusting more than 73 megawatts (250 million Btu per hour) heat input of fossil fuel and providing more than 25 megawatts of electric output to a utility power system. Note that an electric generating unit that does not supply electricity to a utility power system is not covered by this NSPS. (This exemplifies one of the important aspects of NSPS: Their applicability is device-specific. For example, if the electric generating device is part of a combined cycle plant, this NSPS applies only to that portion of the plant burning fossil fuels for steam electric generation and not that part providing steam for heat.)

The NSPS limits are established for particulates, sulfur dioxide, nitrogen oxides, and opacity as shown in Table 6-1. In addition to these emission limits, the NSPS require monitoring and reporting and a detailed compliance testing process. The compliance testing methods are also detailed and must be followed precisely in order to be compliant with the NSPS. Even if the state or other local permitting authority approves a deviation from the NSPS compliance demonstration methods, the same waiver must be obtained from the EPA before compliance with NSPS is assured.

■ TABLE 6-1 **NSPS Emission Limits for Subpart Da Electric Steam Generating Units**

Pollutant	NSPS Limit[a]
Sulfur dioxide—solid fossil fuel	1.20 lb/mmBtu and 90% removal or 70% removal when emissions are less than 0.6 lb/mmBtu
Sulfur dioxide—liquid or gaseous fossil fuel	0.80 lb/mmBtu and 90% removal or 0% removal when emissions are less than 0.2 lb/mmBtu
Nitrogen oxides—solid fossil fuel	0.50 lb/mmBtu subbituminous coal 0.60 lb/mmBtu bituminous coal 0.60 lb/mmBtu anthracite coal 65% removal
Nitrogen oxides—liquid fuel	0.30 lb/mmBtu and 30% removal
Nitrogen oxides—gaseous fuel	0.20 lb/mmBtu and 25% removal
Particulate matter—solid fuel	0.03 lb/mmBtu and 99% removal
Particulate matter—liquid fuel	0.03 lb/mmBtu and 70% removal
Particulate matter—gaseous fuel	0.03 lb/mmBtu
Opacity	Less than 20% except up to 27% for no more than one six-minute period per hour

[a]There are various exemptions and additional requirements for various types of special fuels such as coal-derived liquid or gaseous fuels.

Source: 40 CFR 60.42a–60.44a.

Example NSPS—Hot Asphalt Plants

A second example NSPS is somewhat simpler and involves hot mix asphalt plants (40 CFR Subpart I). This NSPS applies to hot mix asphalt plants constructed or modified after June 11, 1973. The NSPS states that gases discharged from a hot mix asphalt plant shall not contain particulate matter in a concentration greater than 0.04 grains per dry standard cubic foot of exhaust, and the stack gas opacity shall not exceed 20 percent. Compliance testing is simply stated as requiring the standard Method 5 particulate sampling for at least a 60-minute run collecting at least 31.8 dry standard cubic feet of sample volume and using Method 9 for opacity observation.

Example NSPS—Chromium Compounds in Cooling Towers

A third example is the recently promulgated NSPS for chromium compound use in cooling towers. This NSPS simply states that no chromium-containing compound may be used in cooling towers.

These examples illustrate the specificity of NSPS requirements and the range of complexity that may be encountered.

NSPS PROMULGATION SCHEDULE

Prior to the 1990 CAA Amendments, the EPA issued NSPS regulations for only about 40 source categories in the 20 years since the NSPS program was first legislated in 1970. The U.S. Congress considered this rate of promulgation to be less than desirable, and it established in the 1990 CAA an accelerated schedule for promulgation. The 1990 CAA requires the EPA to issue NSPS no later than November 15, 1996, for all source categories initially identified as being possibly subject to NSPS prior to 1990. (NSPS for at least one-half of the source categories must be promulgated by November 15, 1994.)

APPLICABILITY OF NSPS

New Sources

As indicated, NSPS apply to new and modified sources. New sources are those that commence construction after publication of the proposed or final NSPS for that source category. "Commence construction" means that there are actual on-site construction activities on a continuing basis or that the parties have entered into binding contracts to perform on-site construction. In order to be valid contracts, they must not be able to be canceled or modified without substantial loss.

Modified Sources

NSPS also apply to modified sources. A source is considered to be modified, and subject to NSPS, if the following occur:

1. The modification occurs after publication of a proposed or final NSPS.
2. The modification increases the amount of any air pollutant emitted or causes emission of a pollutant not previously emitted.

Modifications do not include the following activities:

1. Routine maintenance activities (such as repair or replacement of equipment)
2. Increase in hours of operation or production (assuming no new equipment is needed and the increases do not violate existing permit conditions)
3. Use of alternative fuels or raw materials
4. Addition or replacement of pollution control technology (unless the new technology is less efficient than the old system)
5. Relocation or change in ownership of the source

Note that if the maintenance activities contemplated in Item 1 above are not ordinary and regular activities necessary to maintain the equipment in its present condition, the EPA states that the activities shall not be considered "routine maintenance," and the modification will be subject to NSPS. For example, if a facility has a 20-year-old gas turbine and wishes to undertake a major overhaul of the turbine to extend its life another 20 years (and the nature of the activities is not of an ongoing nature), the overhaul would be considered a modification and would subject the unit to NSPS.

Reconstructed Sources

Reconstructed sources where the cost of reconstruction exceeds 50 percent of the original fixed capital cost of the source will be considered as new sources for NSPS applicability.

IMPLICATIONS OF NSPS

In general, NSPS is a relatively simple compliance matter for most environmental managers. This is because the NSPS apply on a nationwide basis and are established at levels that have been achieved in practice. Therefore, vendors of equipment can assuredly state that their devices meet NSPS. It is important to note, however, that NSPS provides only the minimum standards that must be met for new sources. The PSD program and state and local regulations often require more stringent limitations than NSPS. It is also important to remember to conduct appropriate compliance testing, routine emissions testing, and recordkeeping as required by the NSPS. In other words, compliance with NSPS includes (a) equipment design and (b) operation and routine ongoing efforts.

On the other hand, NSPS could become a serious problem if a facility is undergoing major overhaul, retrofit, life extension, reconstruction, or other activities that are deemed modifications. In these cases, the existing devices become subject to emission limits established for new devices. Often the existing devices, even with the overhaul and other retrofit activities, cannot meet the more stringent emission limits established for new devices.

NSPS WAIVERS

In order to encourage the use of new and innovative technologies for reducing emissions, a facility can apply to the EPA for a waiver of a specific NSPS in order for the facility to install the innovative technology [Section 111(j)]. The waiver, if granted, expires within 4 years of commencement of construction or 7 years after being granted, whichever is earlier.

In order for a waiver to be granted, the following conditions must be met:

1. The governor of the state in which the source is located must concur with granting the waiver.
2. The proposed system must not have been previously adequately demonstrated.
3. The proposed system will likely achieve greater emission reductions than the NSPS or achieve the same reductions with substantial savings in energy, economic, or non-air-quality environmental impact.
4. The number of waivers granted is not more than necessary to demonstrate for that source category that the innovative technology is achievable.
5. Emissions from the source will not prevent attainment and maintenance of the NAAQS.
6. The waiver applies only to devices for which the innovative technology is being applied.

CHAPTER

7

HAZARDOUS AIR POLLUTANTS

EMISSION STANDARDS FOR HAZARDOUS AIR POLLUTANTS PRIOR TO 1990

In parallel fashion to Section 111 of the CAA, which contains provisions for NSPS, Section 112 of the CAA contains provisions related to establishing National Emission Standards for Hazardous Air Pollutants (NESHAPs). The NESHAPs provisions were contained in the 1970 CAA; but between 1970 and 1990, only eight chemicals were listed as hazardous air pollutants and only 22 emission standards were promulgated prior to January 1993. Frustration by the U.S. Congress over the slow pace of regulating hazardous air pollutants (HAPs) was one of the reasons for the extensive new air toxics program contained in the 1990 CAA.

The primary reason for the slow pace of NESHAP promulgation prior to the 1990 CAA was the basis for HAP listing and NESHAP promulgation established in the 1970 Act. Prior to the 1990 Amendments, the Act required a risk-based approach for listing a HAP. The 1990 Amendments changed that to a technology-based approach.

Prior to the 1990 Amendments, before the EPA could list a chemical as a HAP, the Agency had to conclude that the chemical could reasonably be anticipated to increase mortality or serious illness in humans. Once a chemical was listed as a HAP, the Agency was to promulgate a NESHAP set at levels sufficient to protect the public health with an adequate margin of safety (Blattner, 1992). The major obstacle for the Agency was the scientific uncertainty surrounding health effects from air toxics and the consequent difficulty in developing an administrative record that could withstand court challenge. Therefore, only the following chemicals and NESHAPs were promulgated.

The chemicals that were listed prior to 1990 include

1. Asbestos
2. Benzene
3. Beryllium
4. Coke oven emissions
5. Inorganic arsenic
6. Mercury
7. Radionuclides
8. Vinyl chloride

The NESHAPs promulgated for those chemicals are pollutant- and source-specific. The promulgated NESHAPs, as of January 1993, for these eight chemicals include the following source types:

1. Radon emissions from uranium mines
2. Beryllium emissions in general
3. Beryllium emissions from rocket motor firing
4. Mercury emissions
5. Vinyl chloride emissions
6. Radionuclides from other than Department of Energy (DOE) facilities
7. Radionuclides from facilities licensed by the Nuclear Regulatory Commission
8. Fugitive emissions of benzene
9. Radionuclides from elemental phosphorous plants
10. Benzene emissions from coke by-product recovery plants
11. Asbestos emissions
12. Inorganic arsenic from glass manufacturing plants
13. Inorganic arsenic from primary copper smelters
14. Inorganic arsenic from arsenic trioxide and metallic arsenic production facilities

15. Radon emissions from DOE facilities
16. Radon emissions from phosphogypsum stacks
17. Radon emissions from disposal of uranium mill tailings
18. Fugitive emissions from equipment leaks
19. Radon emissions from operating mill tailings
20. Benzene emissions from benzene storage vessels
21. Benzene emissions from benzene transfer operations
22. Benzene emissions from benzene waste operations

HAZARDOUS AIR POLLUTANT REGULATION AFTER 1990 AMENDMENTS

The 1990 Amendments completely replaced the 1977 Act Section 112 with a new Section 112. The new section retained the 8 HAPs and NESHAPs emission limits (Section 112(q), the "Savings Provision"), but it added 18 new major sections providing for the following:

1. New list of 189 additional HAPs
2. New requirement for Maximum Achievable Control Technology (MACT) for sources emitting HAPs
3. Regulation of both point and area sources of HAPs
4. New requirement to assess residual risk to the public after implementation of MACT
5. Requirement for states to require emission limits to MACT even if EPA has not promulgated the MACT on schedule
6. Program that allows credit for early reduction of HAPs consistent with the 1990 CAA market-based incentive programs
7. New program to prevent and minimize the risk of accidental releases of potentially toxic chemicals
8. New research programs and funding for air toxics research.

Each of these programs will be discussed in the following sections.

LIST OF HAZARDOUS AIR POLLUTANTS

Initial List

As indicated, the U.S. Congress by legislative edict listed 189 chemicals that are deemed to be hazardous air pollutants. The list of 189 chemicals is included in Appendix G and includes common chemicals such as percholoroethylene (dry cleaning solvent), chlorine, benzene (including benzene from gasoline), and 1,1,1-trichloroethane (methyl chloroform or TCA).

Changes to the List

Chemicals can be added to or deleted from the list upon action by the EPA either on its own initiative or after petition from the public. Before a chemical can be added, Section 112(b)(3)(B) requires EPA to determine that the candidate chemical is an air pollutant whose emissions, ambient concentrations, bioaccumulation, or deposition are known to cause or may reasonably be anticipated to cause adverse effects to human health or adverse environmental effects. A chemical can be delisted if the Agency determines that the chemical does not pose the effects noted for adding a chemical. The Agency has 18 months after receipt of a petition to make a decision regarding listing and cannot deny a petition simply for lack of resources necessary to obtain the required data.

The 1990 CAA prohibits the Agency from listing elemental lead as a HAP (lead is regulated through the NAAQS) and also prohibits the PSD program (see Chapter 8) from applying to the HAPs list of chemicals.

There have already been a few petitions for delisting chemicals. In order to successfully file a petition for delisting, the petitioner must obtain the following:

1. Information regarding current peak and annual average emissions of the chemical
2. Information regarding the numbers and types of persons exposed to the chemical
3. An assessment of the health effects of the chemical once released into the environment
4. An assessment of the environmental effects of the chemical

In summary, the petition must provide information to demonstrate that there is adequate data on the health and environmental effects of the substance to determine that emissions, ambient concentrations, bioaccumulation, or deposition of the substance may not reasonably be anticipated to cause any adverse effects to human health or the environment through inhalation or other routes of exposure (58 FR 45081-45082).

From a practical perspective, it will be difficult to *remove* chemicals from the list, but much less difficult to *add* chemicals to the list. The environmental manager should carefully track the Federal Register as to the Agency's progress on petitions to add or delete chemicals, but a reasonable assumption is to assume that all chemicals on the current list will remain on the list.

Proposed Corrections to the List of HAPs

The EPA has reviewed the legislated list of HAPs in detail from a scientific perspective and has proposed certain corrections to the list. The corrections are to further clarify some of the technical issues related to some of the chemicals. Therefore, in Appendix G, the original legislated list and the proposed corrections are both included.

Maximum Achievable Control Technology (MACT) Requirements

Source Categories EPA was required by the Act to list all categories and subcategories of "major sources" and area sources that emit any of the 189 HAPs. This list was required to be published by November 15, 1991 and was actually published in early 1992. Appendix H contains the published list of sources.

"Major sources" are defined, for purposes of the HAPs program, as any stationary source or group of sources within a contiguous area that emit at least 10 tons per year (after controls) of a single HAP or 25 tons per year (after controls) of any combination of HAPs. "Stationary source" is defined the same for HAPs as it is for other new source review (NSR) programs (see Chapter 8).

Note that the term "major source" has a different definition in the HAPs program than in the NSR program (Chapter 8). In fact, the CAA uses the term "major source" in numerous different ways depending upon the specific program. This distinction must be kept carefully in mind when reviewing air quality regulations.

For the HAPs program, the term "area sources" is defined as any source that is not a major source except for motor vehicles or nonroad vehicles subject to regulation by Title II of the Act. This definition is also contrary to common usage in other air quality programs. Normally, the term "area sources" refers to sources of emissions that do not pass through defined vents or stacks or are distributed somewhat evenly over a wide area and are modeled as such in the dispersion models (see Chapter 4). However, in the HAPs program, an area source is simply a widespread collection of smaller point sources.

In the HAPs program, the term "area sources" includes point sources that do have vents or stacks (e.g., gasoline stations, dry cleaning establishments) and even normal manufacturing facilities that do not emit more than 10 or 25 tons per year of a single HAP or combination of HAPs. (For convenience, the 10- and 25-ton-per-year limits for single or combination of HAPs will be referred to in this *Guide* as "10/25 tons per year"). Therefore, the environmental manager must track regulatory developments for "area sources" of HAPs even if his or her facility is considered a point source for all other regulatory programs.

The CAA has several special requirements related to the source category list as follows:

1. Previously regulated source categories can be included in the new list.
2. The source category list was required to include, by November 15, 1995, area sources responsible for at least 90 percent of the 30 HAPs that present the greatest threat to public health.
3. The source category list was required to include, by November 15, 1995, categories and source categories responsible for at least 90 percent of the total emissions of
   a. Alkylated lead compounds
   b. Polycyclic organic matter (POMs)
   c. Hexachlorobenzene
   d. Mercury
   e. Polychlorinated biphenyls (PCBs)
   f. 2,3,7,8-Tetrachlorodibenzofurans (referred to in general as "furans")
   g. 2,3,7,8-Tetrachlorodibenzo-*p*-dioxin (referred to in general as "dioxins")
4. Research or laboratory facilities are to be listed as separate source categories. This is an important distinction, especially with respect to universities, national laboratories, and facilities that conduct product and process development.
5. Boat manufacturing is to be listed as a separate subcategory with respect to styrene emissions.

The significance of the source category list and where a facility falls within the list is that MACT will not apply to a facility that is not one of the source categories, even if the facility emits the same HAP as a facility that was listed. In addition, if a facility is in a separate category or subcategory, it may have a different MACT, again, even if it emits the same HAP as another category. Therefore, it is important to precisely understand the definitions of the categories. Obviously, the Agency wants the categories to be as broad as possible, whereas industry wants the categories to be as narrowly defined as possible. The specific definition of categories and subcategories is one of the first issues addressed by the various working groups established by EPA to develop MACTs.

Definition of MACT

What constitutes MACT is an Agency decision that is source category-specific. However, the Act specifies that MACT must be an emission standard established by the EPA, after taking energy, cost, and non-air-quality health and environmental impacts into account. Contrary to some other programs, the emissions standard that constitutes MACT can be any collection of "measures, processes, methods, systems or techniques," including, but not limited to, one or more of the following [Section 112(d)(2) of the Act]:

1. Volume reduction or product substitution through process changes or other modifications
2. System or process enclosures to eliminate emissions
3. Emission collection and treatment systems (i.e., traditional pollution control systems)
4. Work practices, operational standards, equipment operator and training standards and certifications

As can be seen, the MACT emission limits can greatly affect how a facility conducts its business. Although, where feasible, the MACT standards are to be expressed in terms of numerical limits, MACT is a much farther reaching program than the traditional command-and-control programs currently in effect in most states. This is because it is expected that MACT limits will almost surely include product substitutions and work practice standards. (See, for example, discussion of the Hazardous Organic NESHAP, the "HON Rule," that follows in this chapter.)

MACT applies to both existing and new sources, and the CAA allows the EPA to establish MACT differently for existing and new sources. However, the Act also specifies minimum floors for MACT as follows:

1. For new sources, MACT must be at least as stringent as the best performing new source. This means that MACT for new sources must be at least as stringent as BACT or LAER (see Chapters 8 and 9).
2. For existing sources, MACT must be at least as stringent as the best performing 12 percent of existing sources if a source category has more than 30 sources in it, or the best performing 5 sources if the category has fewer than 30 sources.

One thing to keep in mind is that MACT is promulgated through the NESHAP program. Therefore, as seen in Chapter 10, a MACT source is a NESHAP source and is, therefore, subject to Title V operating permits.

MACT Promulgation Schedule

The CAA specified a rigorous schedule for promulgation of MACT standards. (This was again in frustration over the long time it had taken the EPA to promulgate NESHAPs prior to the 1990 Amendments.) The required schedule for MACT promulgation is as follows:

1. For municipal solid waste incinerators, November 15, 1991
2. For the first 40 source categories, November 15, 1992
3. For coke oven batteries, December 31, 1992
4. For the next 25 percent of source categories, November 15, 1994
5. For industrial waste incinerators, November 15, 1994
6. For publicly owned treatment works (POTWs)—that is, sewage treatment plants—November 15, 1995
7. For the next 25 percent of categories, November 15, 1997
8. For all remaining sources, November 15, 2000

Appendix I lists the final schedule for MACT promulgation that the EPA published in response to the requirements of the Act. As can be seen from Appendix I, the EPA did not list 40 individual categories of sources for MACT to be promulgated by November 15, 1992, but rather included in the first group a major source category, the Synthetic Organic Chemical Manufacturing Industry (SOCMI), which emits a very large number of HAPs and includes a very large number of different kinds of sources. The SOCMI standards for MACT were proposed as the "HON Rule" on December 31, 1992 (57 FR 62608) and promulgated in February 1994. The HON Rule is discussed in detail later in this chapter.

The environmental groups did not accept the SOCMI class as meeting the intent of the Act to promulgate MACT for 40 source categories by November 15, 1992. In response to litigation, the Agency entered into a consent decree to get it on schedule by November 15, 1994. This did not occur either, but the EPA was able to promulgate a few additional MACT standards in 1993 and 1994. Recognizing that the pace is still too slow, in 1995 the EPA announced the MACT partnership program. Under this program, industry, states, local agencies, and environmental groups are invited to submit data regarding existing and potential emission controls. The data are then assimilated and the data become a "presumptive MACT." If no one disagrees with the presumptive MACT, it is promulgated without further review or comment. If there is disagreement, further analysis is conducted until consensus is reached. This process should greatly reduce the time required to promulgate the MACT standard.

MACT Compliance Schedule for Existing Sources

As indicated, MACT applies to both existing and new sources, but only to "major sources" (i.e., sources emitting more than 10/25 tons per year of a HAP).

For new sources that commence construction after promulgation of a MACT standard, the MACT requirements are immediately applicable. The same situation holds for existing sources that undergo "major reconstruction" (as discussed in Chapter 6) such that the source is considered subject to New Source Performance Standards.

For existing sources, the MACT standard will include a compliance schedule for the source category, but can be no longer than three years after promulgation unless

1. The source needs an additional year to install controls and the state agrees to allow the additional year as part of the source's Title V permit (see Chapter 10). (Note that the extension cannot be granted for MACT requirements other than control systems.)
2. Deferrals of any of the MACT provisions can be obtained in 2-year increments (without limit) if the President of the United States determines that the required technology is not available, and such extensions are necessary to preserve the national security interests of the United States. (Each such deferral must be reported to Congress.)
3. An existing source that has installed BACT or LAER (see Chapters 8 and 9) prior to promulgation of a MACT standard shall have a total of 5 years after installation and startup of BACT or LAER to meet the MACT requirements (whether or not the requirements are related to control technology or work practices).
4. If an existing source has undergone a qualifying early reduction in HAP emissions, the source will have a 6-year deferral in complying with the MACT standards. To qualify for an early reduction, the reduction must meet the following requirements:
   a. It should be included as a permit condition under the Title V permitting program.

b. The newly permitted emissions must be less than 10 percent (i.e., at least a 90 percent reduction) of *actual* gaseous emissions of the HAP (95 percent reduction for particulate HAPs).

c. The reductions are measured from actual emissions in a base year chosen by the facility that cannot be earlier than 1985.

d. The reductions (or binding commitments to the reductions) must have occurred prior to the *proposal* of a MACT standard.

Note that it is generally stated that existing sources have 3 years to comply with the MACT, but as is clear from the above, the compliance time can be much shorter or longer depending upon decisions of EPA when MACT is promulgated or special circumstances related to the source. This again emphasizes the need to actively participate in the MACT standard-setting process. In addition, if an existing source is contemplating any changes in processes or systems that affect the amount of HAPs emitted, the early reduction options should be carefully evaluated. Taking advantage of the early reduction deferral program can help stop the "moving target" problem inherent in technology forcing regulatory programs such as MACT.

MACT Compliance for Modified Sources

Consistent with NSPS, modifications to major sources that cause an increase in emissions of HAPs greater than *de minimis* amounts (to be established by EPA) trigger MACT (or an equivalent emission limit established by the state). The practical implication of this provision is that an existing source could be operating without MACT being applied (either because MACT has not been promulgated or the compliance date is not yet passed), but, by virtue of the modification, trigger MACT early.

The CAA specifically allows for netting of HAPs emissions at the facility when determining if the modification causes an increase greater than *de minimis* amounts. This is termed *facility bubbling* and is allowed in the Title V permit program and other programs within the CAA. However, in the case of HAPs, the bubbling is not simply on a mass-to-mass basis (e.g., pounds per hour), but also on a degree-of-hazard basis. In other words, the offset must be for a HAP that is more hazardous than the HAP that is being increased in order to get credit for the reduction.

How to evaluate the degree of hazard and offset rules under the facility bubble is very complicated. Therefore, EPA has yet to issue regulations and *de minimis* values for HAPs.

MACT Compliance for Radionuclides

The EPA will not be required to issue additional MACT requirements for sources of radionuclides, if licensed by the Nuclear Regulatory Commission (NRC). However, if the EPA determines that NRC licensing requirements are not sufficient to protect the public health with an adequate margin of safety, then the EPA can issue additional MACT requirements. In addition, the states can require emission limits more stringent than the NRC if they have the legislative and regulatory authority.

State Programs Equivalent to MACT

States can always implement programs more stringent than federal programs (see Chapter 3). However, the CAA requires states to implement MACT even if EPA misses its deadlines for MACT promulgation.

State Programs for New Sources The Act, under Section 112(g)(2), prohibits states from issuing permits to new, reconstructed, or modified sources (when such modification causes an increase in HAPs emissions greater than *de minimis* amounts) unless those permits contain MACT emission limits (if promulgated) or limits that are equivalent to MACT if MACT has not yet been promulgated. Therefore, under this section of the Act, if EPA does not promulgate a MACT standard, the local permitting authority will be required to make case-by-case determinations as to what constitutes MACT.

State Programs for Existing Sources The provisions for existing sources are established by Section 112(j) of the Act and are termed *equivalent emission limitations by permit.* Proposed regulations for the equivalency program were published by the EPA on July 13, 1993 (58 FR 37778).

The equivalency provisions require a state, once it has an approved Title V permit program in place, to require existing and new sources to meet MACT regardless of the status of federal MACT promulgation. Within 18 months after MACT was to be promulgated by EPA, the states must request all sources within the specific category for which MACT was due to submit a Title V permit application to the state. The state must review the application and issue a permit that requires emission reductions equivalent to those that would have been required if MACT had been promulgated. As an alternative, the state can require a 90 or 95 percent reduction in gaseous or particulate HAPs consistent with the early reduction rule.

The net effect of these provisions is that, no matter what the EPA does with respect to MACT standards, MACT-type limitations will be specified on at least the state level within 18 months of the original congressional schedule.

REGULATION OF AREA SOURCES OF HAZARDOUS AIR POLLUTANTS

MACT applies to major sources. For area sources, the Act allows the EPA to establish Generally Available Control Technology (GACT) standards rather than MACT if it so desires. However, the Agency is not prohibited from requiring MACT on area sources. (Remember that area sources are simply all sources that do not emit more than 10/25 tons per year of HAPs.)

The area source program is detailed in Section 112(k) and requires the EPA to develop, publish, and implement a national strategy of emission reductions from area sources. The strategy must be sufficient to reduce the incidence of cancer attributable to area sources by at least 75 percent.

The Agency is required to conduct a monitoring and research program to develop the area source strategy, report the results of the research to Congress by November 15, 1993, and report the strategy by November 15, 1995. The strategy must at least include GACT (or MACT if necessary) on area source categories sufficient to reduce at least 90 percent of the emissions of the 30 HAPs attributable to area sources that present the greatest risk to public health. The GACT/MACT requirements of the published strategy must be in place on area sources by November 15, 1999.

RESIDUAL RISK AND THE SECOND ROUND OF MACT

One of the more controversial aspects of the new HAPs program contained in the Act is the residual risk assessment process and the possibility of a second round of more stringent MACT requirements. As indicated previously, MACT requirements are technology-driven and are related to health effects only indirectly through the initial HAPs listings and assessment of priorities to establish a schedule for MACT promulgation. The residual risk assessment process, contained in Section 112(f) of the Act, reverses that process and ties the second round of MACT directly to health risk assessments.

Section 112(f) requires the EPA to assess the remaining health risk to the public after installation of MACT on all of the listed source categories. This is termed "residual risk," and the assessment must be delivered to

Congress by November 15, 1996. The report must not only include an assessment of the residual risk and health effects, but also recommend legislation regarding methods to further reduce such risk.

If the Congress does not act on the recommendations presented in the report, the EPA must then issue revised MACT standards for source categories within 8 years of the initial promulgation (e.g., no later than November 15, 2000, for MACT issued on November 15, 1992). The revised MACT standards must be sufficient to protect the public health with an adequate margin of safety. In the case of carcinogenic substances, the revised MACT must at least reduce the risk to less than 1 in 1,000,000 (Chapter 5).

Existing sources will have only 90 days to comply with the revised MACT unless they need a 2-year extension for installation of control technology. New sources constructed or reconstructed after promulgation of the original MACT (i.e., complying with the original MACT), but before promulgation of the revised, more stringent MACT, will not have to comply with the revised MACT until 10 years after promulgation.

The practical impact of this section of the Act is that a facility should evaluate the risk of current facility HAPs emissions using the techniques discussed in Chapter 5. If, after assuming emission reductions of 90 to 95 percent (an assumed minimum value of reduction that would be achieved with original MACT), the screening risk assessment still indicates a risk greater than 1 in 1,000,000 or a hazard index greater than 0.5, the facility should consider eliminating the chemical if possible. This is especially the case for chemicals such as hexavalent chromium, where extremely small amounts still yield an apparent risk over 1 in 1,000,000. If the facility cannot eliminate such a chemical, it is likely that the second round of MACT will require such elimination.

MACT FOR ELECTRIC UTILITIES

Although fossil fuel-fired electric utility boilers and turbines emit many of the HAPs on the congressional list, the CAA specifically exempts utilities pending a special study of utility emissions. The special study was to be submitted to Congress before November 15, 1993, and the report must include a discussion of alternative control strategies for emissions that may warrant regulation under the overall provisions of the HAPs program (i.e., protection of public health).

It is important to note that the electric utility exemption applies only to electrical generation units greater than 25 megawatts capacity or cogeneration units that sell more than one-third and more than 25 megawatts of their electrical capacity to a power distribution system.

EXAMPLE MACT—THE "HON RULE"

On December 31, 1992, the EPA proposed the Hazardous Organic NESHAP rule, called the HON Rule. The final rule was promulgated in February 1994. This rule is the MACT for one source category, the Synthetic Organic Chemical Manufacturing Industry (SOCMI), and seven additional processes in other source categories. It is an extremely complicated rule that includes the concepts of

1. Facility bubbling
2. Emission averaging
3. Work practice standards
4. Resource allocation allowing facilities to choose the most efficient application of controls

The rule required 76 pages of the Federal Register to explain and 137 pages of Federal Register regulatory language. Examination of this rule is illustrative for understanding future MACT standards.

Applicability

One of the first concepts of the rule is applicability. Applicability is determined in a three-step process. In order for the rule to apply to a facility, the facility must produce one or more of a list of 396 synthetic organic chemicals. Secondly, the facilities that produce one or more of the 396 synthetic organic chemicals must use as a reactant, or produce as a product, co-product, intermediate, or by-product, one or more of a list of 149 organic hazardous air pollutants (HAPs) normally associated with SOCMI. (Both the lists of synthetic organic chemicals and organic HAPs are contained in the rule.) Thirdly, the facility must emit more than 10 tons per year of a single HAP or 25 tons per year total HAPs. If the facility does not meet all three conditions, the HON rule does not apply.

Emission Limit Requirements

The HON rule establishes requirements for four types of emission points plus equipment leaks for SOCMI facilities. It also contains an "equipment leaks only" rule for the seven non-SOCMI processes and chemicals affected by the rule. The non-SOCMI processes and chemicals are

1. Styrene–butadiene rubber production (butadiene, styrene)
2. Polybutadiene production (butadiene)
3. Chlorine production (carbon tetrachloride)
4. Pesticide production (carbon tetrachloride, methylene chloride, ethylene dichloride)
5. Chlorinated hydrocarbon use (tetrachloroethylene, chloroform, carbon tetrachloride, methylene chloride, ethylene dichloride)
6. Pharmaceutical production (carbon tetrachloride, methylene chloride)
7. Miscellaneous butadiene use (butadiene)

The five types of emission points for the SOCMI sources are

1. Process vents
2. Storage vessels
3. Transfer operations
4. Wastewater streams and associated residuals
5. Equipment leaks

For the SOCMI facilities, the rule requires the facility to reach an overall facility emission limit *plus* minimum control efficiencies on a subset (called Group 1) of three of the five types of emitting points (vents, vessels, and transfer). The Group 1 emitting points are those that have the least resource requirements for installation of controls and/or the greatest HAP emissions. There are various conditions that place an emitting device in Group 1, but one of these is calculation of a Total Resource Effectiveness (TRE) index. Devices that are not Group 1 are Group 2 devices.

The overall facility emission limit is calculated assuming that Group 1 process events should have an average control of 98 percent; Group 1 storage vessels an average of 95 percent control; Group 1 transfer racks an average of 98 percent control; and Group 1 wastewater operations various device-specific control requirements.

Once the overall facility limit is established for the first four types of emitting devices (i.e., leaking equipment is not included in the facility emission limit), the facility is allowed to control whatever devices (including Group 2 devices) it chooses as long as the overall limit is met. However, if the facility chooses not to control a certain Group 1 device, there is a penalty (currently proposed as 20 percent) on the emissions from that device. In other words, if, for example, a Group 2 device is controlled instead of a Group 1 device, the control on the Group 2 device must remove the same amount of emissions as the Group 1 device control plus 20 percent additional removal.

For equipment leaks, facility bubbling is not allowed, because the EPA states that no accurate method exists for calculating the emissions from leaking equipment nor for establishing control efficiencies. Therefore, the rule only requires various inspections, repair requirements, and monitoring in order to minimize leaks as much as possible. In this manner, the equipment leak requirement is much more of a work practice standard than an emission limit. Such a combination of work practice requirements and emission limits should be expected in nearly all of the MACTs to be promulgated by the EPA.

Reporting Requirements

The HON rule also requires extensive recordkeeping and reporting. The first report is an initial notification that lists the chemical manufacturing processes at the facility that are subject to the rule and which provisions apply. The notice also requires a statement by the facility that it can achieve compliance by the required dates.

The second required report is the implementation plan, which details how the facility intends to come into compliance. This plan will include details regarding the group classification of facility devices and the facility's proposed bubbling plan, if any. The third report is a notification of compliance status, which contains the necessary information to demonstrate compliance. The fourth reporting requirement is submittal of periodic emission reports, including quarterly reports for all points included in an emissions average. The fifth reporting requirement is to submit other special reports to reflect various conditions (e.g., a report of startup, shutdown, and malfunction events).

HON Rule Summary

The HON rule is a useful indicator of the future with respect to the types of MACT standards that will be issued. In future MACT standards, there will likely be combinations of emission limits, facility operational flexibility to meet those limits, and various work practice standards. For larger facilities, these requirements will be tied together with extensive reporting requirements, including implementation plans and compliance verification reports.

EXAMPLE MACT—ASBESTOS

The asbestos NESHAP is a good example of a work practice type of MACT standard. It is promulgated in 40 CFR 61.140-156 and applies to manufacturing, milling, spraying, fabricating, insulating, and disposing of asbestos. It also applies to demolition and renovation operations involving friable asbestos material and roads that may contain asbestos materials.

The demolition and renovation restrictions are promulgated in 40 CFR 61.145 and apply to demolition and renovation activities where friable asbestos can potentially be released to the atmosphere. Friable asbestos is material that contains more than 1 percent asbestos by weight and that can crumble, pulverize, or reduce to powder when dry. In addition, if there is asbestos in roof products and floor tiles and the demolition activities involve cutting or sanding of such materials, the asbestos in them is considered friable.

The manufacturing and milling aspects of the NESHAP are traditional emission control systems. The NESHAP requires that no visible emissions occur and also requires that effective air filters that capture the asbestos fibers be placed on devices that could potentially emit asbestos. The disposal requirements of the NESHAP specify operational requirements that notify the public that asbestos has been disposed of at the site and that prevent the asbestos from being re-entrained into the environment after disposal.

The demolition and renovation NESHAP requires prior advance notification to the EPA and local agencies that demolition activities involving friable asbestos are about to occur. In addition, work practices must be implemented to reduce the possibility of emitting friable asbestos into the atmosphere or exposing workers to asbestos fibers. These activities include removal of the asbestos prior to wrecking or other demolition activities, wetting the asbestos-containing materials, and sealing the materials in plastic bags, among other practices. When removed, the asbestos must be manifested and transported to an approved and permitted disposal site that meets the NESHAP requirements for disposal.

It is easy to see from the asbestos example how comprehensive the various MACT standards will be. These standards will almost certainly require much greater involvement and modification to existing work practices than any other air quality regulations. For example, in the asbestos NESHAP, there are at least 10 detailed paragraphs dealing only with the size and type of labels that must be placed at facilities handling asbestos (including the font type that is allowed—only sans serif, gothic, or block fonts are allowed).

REGULATIONS RELATED TO ACCIDENTAL RELEASES OF TOXIC CHEMICALS

For the first time in history, the CAA regulates potential *accidental* releases of toxic chemicals as well as *routine* releases. At least two states, New Jersey and California, already have such risk management and prevention plan legislation and regulations, and the federal program will be tailored after these states. Therefore, in anticipation of the federal program, consulting these two state programs will be very useful.

The federal accidental release program requirements are detailed in Section 112(r) of the Act and require the following steps:

1. Publicize a list of chemicals that potentially pose serious adverse health effects when released.
2. Establish threshold quantities of each chemical listed.
3. Require registration of facilities that handle, store, or produce more than the threshold quantities.
4. Promulgate regulations to require affected facilities to prevent and detect accidental releases, provide adequate response to releases, and develop and implement risk management plans.

Section 112(r)(6) also establishes a Chemical Safety and Hazard Investigation Board to investigate accidental releases resulting in fatality, serious injury, or substantial property damage. The Board is also required to advise the EPA on risk management and hazard assessment methods to reduce the possibility and potential severity of releases.

List of Chemicals

The list of chemicals deemed by the EPA to pose a serious threat to health if released was proposed on January 19, 1993 (58 FR 5102). This list of chemicals is reproduced in Appendix J and includes 100 acutely toxic chemicals, 62 flammable gases and liquids, and the U.S. Department of Transportation (DOT) list of high explosives. This list of chemicals includes some very common chemicals such as ammonia, chlorine, sulfur dioxide, and toluene diisocyanate (used in foam production).

Risk Management and Prevention Plan Elements

Proposed regulations were published by the EPA on October 29, 1993, requiring risk management plans to include at least:

1. A hazard assessment to evaluate the potential for accidental releases
2. A program for preventing the releases, including engineering design, maintenance procedures, monitoring, and employee training
3. A program for responding to releases

The regulations must cover handling, storage, production, maintenance, repair and replacement activities, and detection of releases. Facilities that have more than the threshold quantities of the listed chemicals must implement the regulations no later than 3 years after promulgation (i.e., anticipated to be no later than November 15, 1999) or 3 years after the facility first handles quantities above the threshold quantities of the chemical, whichever is later.

Two major states, California and New Jersey, have formal risk management programs. The California program requires affected facilities to submit a Risk Management and Prevention Plan (RMPP) for review and approval by the local administering agency (the local Fire Departments or County Hazardous Materials Management Departments). The elements of an RMPP include the following:

1. *Hazard and Operability (HazOp) Study.* The HazOp is a formal process of reviewing the design and operational procedures at the facility in order to identify potential mishaps that could cause a release.
2. *Seismic Analysis.* The seismic analysis identifies the most probable severe earthquake and, through engineering principles and calculations, evaluates chemical release scenarios resulting from seismic events.
3. *External Events Analysis.* The external events analysis evaluates the possibility of chemical release as the result of external events (e.g., airplane crash into the facility) other than seismic or facility fire. (A facility fire is evaluated under the HazOp.)
4. *Offsite Consequence Analysis.* This analysis uses atmospheric dispersion modeling to evaluate the concentrations at downwind locations of pollutants released in a mishap. The modeled concentrations are compared to levels of concern (LOC) established by the local agency or through toxicological evaluation.
5. *RMPP Document.* The technical studies listed in items 1 through 5 are generally kept confidential to the agency, because they reveal details regarding the facility that could be considered trade secrets or are otherwise sensitive. On the other hand, the RMPP document is a compilation and summary of the technical studies and is made public. The document also identifies management plans that have been or will be undertaken to minimize the risk and severity of an accidental release.

Practical Use of the RMPP

The RMPP is basically sound management and engineering. Contrary to some regulatory programs, preparing and implementing an RMPP at a facility will actually benefit the facility by improving facility operations and procedures.

Similarly, facilities that are undergoing modifications or engineering upgrades and that handle more than the threshold quantities of the listed chemicals should conduct hazard evaluations now, even before the regulations are promulgated and implemented. The methods for such evaluations are well known and available from several sources (e.g., AICHE, 1992). By conducting such reviews now, engineering and operational safeguards can be built into the system and expensive retrofits may be avoided.

CHAPTER 8

NEW SOURCE REVIEW

FUNDAMENTAL CONCEPTS

Whenever a new source of emissions is contemplated, the owner of the new source will have to proceed through some type of New Source Review (NSR) process before it can be legally constructed. This process can be as simple as determining that the emissions are below permitting thresholds, and therefore the source can be constructed at will with no regulatory oversight. On the other hand, the process can be extremely complicated involving numerous federal, state, and local agencies and requiring years to complete.

There are two excellent references available that describe in detail the NSR process. One is the *New Source Review Workshop Manual* published by the EPA in October 1990. This manual is available by contacting the local EPA regional office and paying a copying charge to obtain a photocopy. A proposed revision to the *Manual* was published through the EPA TTN bulletin board in July 1994, and a second proposed revision was published in late 1995. The second reference is a three-volume compilation of EPA policy memoranda and other interpretive information available from the Air & Waste Management Association (headquarters located in Pittsburgh, Pennsylvania) and is entitled the *New Source Review Prevention of Significant Deterioration and Nonattainment Area Guidance Notebook* (AWMA, 1991). This compilation is very useful because it addresses some of the specific, detailed, and somewhat arcane policies related to the more complex NSR issues.

It is important to realize that, just as New Source Performance Standards (Chapter 6) apply to modifications of existing facilities as well as new sources, the NSR process also applies to modifications of existing facilities that increase emissions by more than *de minimis* amounts.

The NSR process is different for two categories of sources: major and minor. It also differs depending on the two types of areas in which the source is to be built: attainment and nonattainment. Accordingly, there are four different types of NSR processes:

1. Major source in attainment areas, termed *prevention of significant deterioration* (PSD)
2. Major source in nonattainment areas, termed *nonattainment area permitting* (NAA)
3. Minor source in attainment areas, termed *minor source permitting*
4. Minor source in nonattainment areas, termed *minor source nonattainment permitting*

In practice, the two minor source permitting processes are usually grouped together and, for minor sources, there is no difference between attainment and nonattainment minor source permitting.

MAJOR SOURCE DEFINITIONS

In the CAA, the term *major source* is used in numerous different sections, and the term does not always mean the same thing. For NSR purposes, a major source is one that has the "potential to emit" criteria pollutants in amounts greater than certain thresholds (termed *major source thresholds*). The magnitude of the major source thresholds depends on whether the source is in an attainment or nonattainment area and the severity of the nonattainment status of a nonattainment area.

Major Source Thresholds in Attainment Areas

For attainment areas (which include unclassified areas—see Chapter 9), the major source thresholds are potential emissions greater than 250 tons per year of any one criteria pollutant, unless the source is one of the 28 major emitting facilities (MEFs). Pursuant to Section 169(1) of the federal CAA, a source is an MEF if its potential to emit is 100 tons per year or more of any pollutant and it is one of the following 28 major source categories:

1. Fossil-fuel fired steam electric plants of more than 250 million Btu per hour heat input
2. Coal cleaning plants (thermal dryers)
3. Kraft pulp mills
4. Portland cement plants
5. Primary zinc smelters
6. Iron and steel mill plants
7. Primary aluminum ore reduction plants
8. Primary copper smelters
9. Municipal incinerators capable of charging more than 50 tons of refuse per day
10. Hydrofluoric acid plants
11. Sulfuric acid plants
12. Nitric acid plants
13. Petroleum refineries
14. Lime plants
15. Phosphate rock processing plants
16. Coke oven batteries
17. Sulfur recovery plants
18. Carbon black plants (furnace process)
19. Primary lead smelters
20. Fuel conversion plants
21. Sintering plants
22. Secondary metal production facilities
23. Chemical process plants
24. Fossil-fuel boilers of more than 250 million Btu per hour heat input
25. Petroleum storage and transfer facilities with a capacity exceeding 300,000 barrels
26. Taconite ore processing facilities
27. Glass fiber processing plants
28. Charcoal production facilities

In attainment areas, a modification is considered a major modification, and subject to major NSR review, if the modification has the potential to increase the emissions of criteria pollutants at an existing *major* (only) source by more than any one of the following thresholds (termed significant emission increases):

1. Carbon monoxide—100 tons per year
2. Nitrogen oxides—40 tons per year
3. Sulfur oxides—40 tons per year
4. Particulate matter—25 tons per year TSP or 15 tons per year of PM_{10}
5. Volatile organic compounds (VOCs)—40 tons per year
6. Lead—0.6 tons per year

For the noncriteria pollutants, the thresholds vary and can be as small as "any amount" for some pollutants if there is an NSPS established for that pollutant. These thresholds are as follows:

1. Asbestos—0.007 ton per year
2. Beryllium—0.0004 ton per year
3. Mercury—0.1 ton per year
4. Vinyl chloride—1 ton per year
5. Fluorides—3 tons per year
6. Sulfuric acid mist—7 tons per year
7. Hydrogen sulfide (H_2S)—10 tons per year
8. Total reduced sulfur (TRS, including H_2S)—10 tons per year
9. Sulfur compounds (including H_2S)—10 tons per year
10. Radon-222; polonium-210; CFCs 11, 12, 112, 114, 115; HCFCs 1211, 1301, 2402; and any other pollutants regulated by the CAA through NSPS, NESHAPs, Ozone Depletion (Chapter 12), or any other CAA programs. However, per the 1990 CAA, emissions of the 189 federal HAPs *do not* trigger NSR.
11. Any amount of pollutant that could cause greater than 1-microgram-per-cubic-meter impact on a Class I area, or when the source is located within 10 kilometers of a Class I area.

Note that these modification threshold values apply only if the source is major prior to the modification. For example, if a source emits 220 tons per year of VOC and the modification is an increase of 50 tons, the modification would not be a major modification, even though the resulting total is 270 tons, over the new major source threshold. If there is another modification at that facility increasing VOCs by more than 40 tons per year, the second modification would be major, and the entire source would then become subject to NSR as a major source. If the modification is itself "major"—that is, over 100 or 250 tons per year—the entire source would also be considered major.

Internal Offsets and Contemporaneous Reductions

Often, during facility modifications, there is an increase in emissions from one type of device (for example, a new, larger boiler), but there is also a concomitant emission reduction from another device (for example, the older boiler that was replaced). When evaluating the modification threshold for NSR, it is the *net* emissions increase that is evaluated, not the emissions of the proposed device alone. (Note, however, that the new device may still be subject to other regulatory requirements—for example, NSPS, even if NSR does not apply.) This is termed *emissions netting* or *internal offsets*. (External offsets—that is, emission offsets between the proposed source and a separate source—occur in the nonattainment program; see Chapter 9.)

In order for emissions reductions to be creditable in internal netting, the reductions must be:

1. *Contemporaneous*. The emissions reductions must have occurred at the same time as the increase (as in the case of replacement equipment) or within the past five-year period.
2. *Real*. Only *actual* emissions reductions, not permitted, can be counted. This is a common problem in offsetting because the new device must be *permitted* at full capacity and usage. However, rarely does a device *operate* at full capacity and emissions. Experience has shown that actual emissions are usually about one-half the permitted emissions. Therefore, there is a *de facto* 2-to-1 offsetting ratio (i.e., twice as many permitted emission reductions must occur than permitted increases).
3. *Quantifiable*. This means that records and other proof of the magnitude of the emissions must be available. It also means that normal, fugitive emissions, which are generally deemed to not be quantifiable, do not count. Environmental managers should keep good fuel usage data, production rate data, source emissions test results, and other information on file that can be used to determine the actual emission rates of facility devices. These data are useful in a wide variety of air quality settings. (Interestingly, however, it is this author's experience that many facilities do not keep those data available.)

4. *Permanent.* The emissions reductions cannot be temporary, and they must be enforceable. This generally means that one must surrender the permit or accept a permit modification that reduces emissions for the device that is being used to obtain an emissions reduction.
5. *Creditable.* The changes must not have been taken into account when issuing a previous permit or modification for the source.

Only emissions reductions that meet *all* five of the above tests can be used for netting. Note that these same types of requirements also apply to external offset generation (i.e., emission reduction credits) as discussed in Chapter 9.

There is a special exemption for condition 2 above, related to comparing "previous actuals" to "future potential" emissions. This is the like-kind replacement (also often called the *functionally equivalent replacement*) provision. If the new device is a like-kind replacement for old equipment, then one can compare previous actuals to *estimated* future potential emissions (i.e., not the maximum). This exemption makes intuitive sense and allows for a like-kind replacement without triggering NSR, but it has been the subject of much litigation.

Major Source Thresholds in Nonattainment Areas

In nonattainment areas, the major source thresholds range from 10 tons per year to 100 tons per year depending on the pollutant and the severity of the area's nonattainment status. The various classifications of nonattainment areas are discussed in Chapter 9. The thresholds for new sources in those areas are shown in Table 8-1.

Major modifications in nonattainment areas also have different thresholds than attainment areas, and those thresholds are shown in Table 8-2.

Local Definitions of Major Source

Some local permitting jurisdictions have lower major source or major modification thresholds than those federally required and shown in Tables 8-1 and 8.2. For example, in the Los Angeles area, although the federally required major source threshold is 10 tons per year for VOCs (because it is an extreme nonattainment area), the local permitting authority has established a zero threshold. In other words, in the Los Angeles area, all sources are considered "major sources."

■ **TABLE 8-1** Major Source Thresholds in Attainment and Nonattainment Areas (Tons per Year)

Pollutant	Attainment Area	Nonattainment Area Status					
		None	Marginal	Moderate	Serious	Severe	Extreme
VOCs (ozone)	100/250[a]	—	100	100	50	25	10
Carbon monoxide	100/250[a]	—	NA[c]	100	50	NA	NA
PM_{10}	100/250[a]	—	NA[c]	100	70	NA	NA
Nitrogen oxides	100/250[a]	100[b]	NA	NA	NA	NA	NA
Sulfur oxides	100/250[a]	100[b]	NA	NA	NA	NA	NA

[a]100 TPY for major emitting facility, 250 TPY for other sources.
[b]There are no classification status designations for nitrogen or sulfur oxide nonattainment areas; all nonattainment areas are treated the same.
[c]PM_{10} and CO nonattainment areas are classified either as moderate or serious. NA, not applicable.

■ **TABLE 8-2** Major Modification Thresholds in Attainment and Nonattainment Areas (Tons per Year)

Pollutant	Attainment Area	Nonattainment Area
VOCs (ozone)		
Marginal	40	40
Moderate	40	40
Serious	40	25
Severe	40	25
Extreme	40	10
Carbon monoxide	100	100
TSP/PM_{10}	25/15	25/15
Nitrogen oxides	40	40
Sulfur oxides	40	40

This situation is due to the severity of the nonattainment status of the area as well as the fact that the local permitting authority permits all sources as if they were major and then allows various exemptions and regulatory relief for some smaller sources. Most other permitting authorities handle major sources and minor sources differently and specify different permitting processes for these two different types of sources.

MINOR SOURCE DEFINITIONS

Any new or modified source that has a potential to emit or potential to increase emissions of a pollutant that is not a major source is a "minor source." If a source is minor, it is reviewed and permitted under the various minor source provisions of the local permitting authority. There are no purely federal permitting requirements for minor sources, but there are almost always local permitting requirements for minor sources.

The local permitting authorities have a wide variety of permitting systems for minor sources. However, these systems can generally be classified into two types.

One type of local permitting system is to require permits from sources that have emissions greater than a *de minimis* amount—for example, 10 pounds per day (or 1.3 to 1.8 tons per year, depending on 260-day or 365-day-per-year operations) of criteria pollutants or pollutants regulated by the local permitting authority. In this system, if a source emits less than the *de minimis* amount, no permit is required, and the source can be constructed without regulatory oversight.

The other type of permitting system, which tends to be used throughout California, is to require permits of *all* sources that emit *any amount* of *any air contaminant*, unless the source is exempted by rule from needing a permit. This system tends to permit thousands of sources per year, because the exemptions are very narrowly crafted. (For example, a typical exemption is to exempt degreasers that are not heated, covered when not in use, and have an open area when in use of less than 1 square foot.)

POTENTIAL TO EMIT

One of the key considerations as to whether a source is major or minor is the potential to emit. All of the various thresholds are based on the potential to emit.

The potential to emit is that quantity of emissions that would occur under maximum design capacity except as constrained by federally enforceable conditions (which generally include the effects of emissions controls and/or restrictions on hours of operation, or the type of material combusted, stored, or processed). Simply stated, the potential to emit is the amount of emissions "after controls." However, the key condition regarding controls is that they must be related to a federally enforceable condition.

For example, if a major new source is to be built, it must install at least best available control technology (BACT), which is discussed in the following sections. If, for example, the BACT system reduces emissions by 95 percent, the potential to emit would be only 5 percent of the uncontrolled emissions.

The question becomes, What if the source appears to be major without controls but minor with controls? This would occur, for example, if the uncontrolled emissions are 300 tons per year, but the controls reduce emissions to only 15 tons per year (95 percent removal, for example). In this case, if the controls were installed and operated pursuant to a federally enforceable condition, the source would be a minor source.

Federally enforceable conditions are those restrictions that are enforced by the local permitting authority through a federally approved State Implementation Plan (SIP, Chapter 3) or by the EPA through its permitting system. In general, any limit on emissions that the facility is willing to accept can be considered a federally enforceable condition. However, there is a special case when the proposed facility is located on Native American land.

On Native American land, unless the Native American tribe has submitted an approved Tribal Implementation Plan (TIP), the CAA is enforced by the EPA. As stated above, the EPA has no minor source permitting program (in contrast to the states that do have a minor source permitting program). Therefore, it is the position of the EPA that no system of controls that moves an apparent major source to a minor source category can be considered federally enforceable on Native American land unless the tribe has an approved TIP (see Chapter 3). (No Native American tribe has yet to receive an approved TIP.)

In the case of Native American lands when there is not an approved TIP, the potential to emit is considered to be the emissions without controls. This is because if controls were included, the source would be minor and the EPA would have no jurisdiction—that is, no federally enforceable limitation. This seemingly bizarre interpretation and application of the definition of potential to emit illustrates the importance of the concept of "federally enforceable" in the CAA program.

DEFINITION OF SOURCE

A second key definition is what constitutes a "source." The NSR program applies to stationary sources only. When the facility consists of a single point of emissions, it is fairly obvious that the source is the facility and its single emission point. Even when the facility consists of several emitting points from a single building, it is fairly obvious that the source is all of the emitting points from that building. However, when the facility consists of a number of different buildings and operations, the definition of source becomes more complicated.

The term "source" is defined as a building, structure, facility, or installation that meets all of the following criteria. The building, structure, facility, or installation must

1. Belong to the same industrial grouping
2. Be located on one or more adjacent or contiguous properties
3. Be under common control or ownership

Industrial Grouping

The first criterion, same industrial grouping, is evaluated by determining whether the operations at the installation have the same two-digit SIC (Standard Industrial Classification) code. The SIC code system is established by the Federal Office of Management and Budget in a book published by the Government Printing Office (USOMB, 1987).

The SIC codes consist of a total of four digits, with the first two digits establishing a major group of industrial activity—for example, "primary metal industries." The third digit establishes an industry subgroup—for example, "steel works, blast furnaces, and rolling and finishing mills." The fourth digit further specifies the industry—for example, "steel works, blast furnaces (including coke ovens), and rolling mills." For reference, Appendix K contains a list of two-digit SIC codes and the associated major industry grouping description. The SIC code for a facility is usually established by the facility accounting department, because SIC codes are routinely used in financial reporting. However, the choice of SIC codes is determined by the facility, not by the EPA, and the choice of SIC codes can have far-reaching consequences.

If two operations have different two-digit SIC codes, for example, one operation is a metal foundry, and the other is an automobile manufacturer; the two operations cannot be considered the same source, and the emissions from the two operations are *not* summed when determining the potential to emit.

Contiguous or Adjacent Property

The second criterion of source deals with location of the operations. If the properties are adjacent (i.e., touching) and have the same two-digit SIC code and are under common control and ownership, the facility is a single source. The question is, What if the properties are nearby, but not touching (i.e., contiguous)? Under the EPA definition, if the properties are contiguous but not adjacent, the operations could still be considered a single source. The EPA has not established a definition of "contiguous," but many local permitting authorities have stated that if the properties are separated by no more than a public roadway, they are contiguous. If there is more than a public roadway between the properties (e.g., a lot owned by a third party), they would not be contiguous. (Some other authorities use the concept of the ability to trace a single line around the properties on which the emitting devices are located as indicative of the source being contiguous.)

However, the EPA does not always use this common interpretation of contiguous. For example, the EPA has considered compressor stations separated by 20 to 100 miles, but connected by a pipeline, as being contiguous sources. Therefore, if the facility includes disbursed but connected sources, and the other two criteria are met, consultation with the Agency is probably necessary to resolve the "contiguous" issue.

Common Control and Ownership

The third criterion is common control and ownership. This criterion is apparently relatively simple, but can become complicated in some situations (e.g., federal installations and subsidiary operations). As in the other criteria, the EPA has not defined what is meant by common control and ownership. Usually, the intuitive definition is used. For example, in the case of federal facilities, if the installation is under a single agency, let us say the U.S. Navy, it is a single source, even if the installation houses multiple tenant commands within the overall agency. However, if the installation houses multiple agencies (e.g., the U.S. Navy, the Department of Energy, and the Federal Aviation Administration), then it is possible that the EPA would *not* consider all tenants to be a single source.

For private companies, it is generally considered that parent and subsidiary operations are under common control and ownership. However, if the subsidiary is completely independent from the parent with respect to decisions regarding air pollutant emissions, it may be possible to consider the parent and subsidiary as not being under common control and ownership for purposes of NSR review.

TYPES OF EMISSIONS

When the source is defined, there is a question as to what emissions are included in the "potential to emit" calculation. The different potential emission types are point source emissions, fugitive emissions, mobile source emissions, and secondary emissions.

Point source emissions are those that pass through a vent, stack, chimney, or functionally equivalent opening that is stationary. Fugitive emissions are those that do not pass (or could not reasonably pass) through a vent, stack, chimney, or other functionally equivalent opening.

Mobile source emissions are emissions from vehicles capable of being licensed to travel on a public roadway. Note, therefore, that mobile source emissions do not include emissions from forklifts or construction equipment. These types of emissions would be considered fugitive emissions. Note also that if emissions could *reasonably* pass through a vent or stack, even if they do not under the proposed design, the emissions could be considered point emissions rather than fugitives. This would occur, for example, in a situation where forklifts were used in an enclosed warehouse that, due to poor facility design, was vented only through open doorways and windows. Clearly, a venting system could (and should) be installed, and the forklift emissions would, in this case, be considered as part of the point source emissions.

Secondary emissions are those emissions that would not otherwise occur except for the fact that the major source is to be built. Secondary emissions do not include emissions from mobile sources, as defined above. (Note that secondary emissions could include emissions from construction equipment or other off-road vehicles. Also note that emissions from vessels at berth are not included as secondary emissions, but are included as primary emissions associated with the major source itself.) An example of secondary emissions would be the additional emissions occurring at a limestone quarry associated with a cement plant expansion.

For purposes of determining the potential to emit and major source thresholds, point source emissions always count, secondary emissions do not count, mobile source emissions do not count, and fugitive emissions count only if the source is one of the 28 major emitting facility source categories.

Even if various types of emissions (i.e., mobile, secondary, and fugitive under some circumstances) are not counted in determining major source threshold applicability, they must be included as part of the ambient air quality impact evaluation of the source. (Note, however, that secondary emissions must be specific, well-defined, and quantifiable and must affect the same general area as the stationary source or modification undergoing review in order to be included in the impact evaluation.)

PREVENTION OF SIGNIFICANT DETERIORATION PROGRAM

When it is determined that a source is a major source and will be locating in an attainment area, it is subject to the Prevention of Significant Deterioration (PSD) program. (If the source is locating in a nonattainment area, it is subject to the nonattainment NSR program discussed in Chapter 9.)

Note that a source is subject to the PSD program *only* for those pollutants that are "major" with respect to the major source thresholds and only if those major pollutants are attainment. For a source located in a developing area that emits varying amounts of different pollutants, which is the normal case, it is likely that the source will have to undergo a combination of NSR programs.

For example, assume that a source emits 270 tons of sulfur dioxide per year, 70 tons per year of VOCs, and 150 tons per year of particulate. Also assume that the source is locating in an area that is attainment for sulfur dioxide and particulate, but nonattainment for ozone. In this case, the source will undergo nonattainment area (NAA) permitting for VOCs (because it is major under the nonattainment program for ozone), PSD permitting for sulfur oxides (because it is major for sulfur oxides), and minor source permitting for particulate (because it is minor for particulate emissions). If, however, the area was also nonattainment for particulate, the major source threshold drops to 100 tons per year for particulate, and the source would have to undergo NAA review for particulate also, because it would be major for particulate in a nonattainment area. Obviously, it can be quite confusing, and the individual pollutants from a source must be as carefully examined as the total emissions of the source.

The PSD program requires the source to conduct a number of analyses and demonstrations for point and fugitive emissions from the source, as well as potential air quality impacts related to additional growth associated with the source. The required elements are as follows:

1. Conduct preconstruction ambient air quality and meteorological monitoring and possible postconstruction monitoring.

2. Conduct an ambient air quality impact analysis (AQIA) to demonstrate that the NAAQS are not exceeded.
3. Conduct an ambient air quality impact analysis (AQIA) to demonstrate that the PSD increments are not exceeded.
4. Demonstrate that the source will not impair visibility or adversely affect soils or vegetation in the impact region of the source.
5. Demonstrate that the source will not adversely affect air quality related values (AQRVs), including visibility, at Class I areas (national parks and major wilderness areas).
6. Demonstrate that the source will be constructed with Best Available Control Technology (BACT) regardless of the magnitude of the ambient air quality impact.
7. Allow for significant public participation and review of the proposed project prior to the EPA taking final action on the permit.

Baseline Date

One of the precepts of the PSD program is that the air quality in a region cannot become significantly worse than it was when the PSD program was first established for a region. This date is termed the *baseline date*, and the ambient air quality that existed on the baseline date is termed the *baseline concentration* (which is established for each pollutant). The region for which the baseline date is established is termed the "baseline area."

The baseline date is triggered in an area when the first complete PSD application is filed for that region. The source for which the PSD application was filed is called the *baseline trigger source.* Since the PSD program was legislated in the 1977 Clean Air Act Amendments, the baseline date can be no earlier than August 7, 1977 for particulate matter (PM) and SO_2 and no earlier than February 8, 1988 for NO_2 (the NO_2 date is later because, initially, only SO_2 and PM were in the program, and NO_2 was not added until 1988).

In practice, most areas that experienced normal growth have baseline dates in the 1978 to 1980 time frame for PM and SO_2 and 1989 to 1990 for NO_2. Note that the baseline date is established by air quality control region (AQCR; see Chapter 3). The formal baseline dates are available from the local EPA region or local permitting authority.

The baseline area is the AQCR in which the proposed source is located plus any AQCR in which the trigger source had an ambient air quality impact greater than 1 microgram per cubic meter (annual average). If necessary, a state can redefine the baseline area to areas less than the AQCR as long as the area is at least as great as the 1-microgram-per-cubic-meter impact zone of a baseline trigger source.

A circle around the source whose radius is the greatest distance to a point where the ambient air quality impact is at least equal to 1 microgram per cubic meter (annual average) is termed the *impact zone* of the source. In order to determine the impact zone, ambient air quality modeling (Chapter 4) must be conducted. The impact zone concept is used throughout the PSD program to define the area in which various analyses must be conducted. Because the impact zone is a circular area, it is also sometimes referred to as the "impact circle."

Ambient Air Quality and Meteorological Monitoring

In order to determine the air quality that exists in a region prior to construction of the proposed major source, the source owner is required as part of the PSD program to conduct baseline monitoring for each of the criteria pollutants that will be emitted by the source. The source can obtain an exemption from such monitoring if it can be demonstrated that the source will cause an insignificant ambient air quality impact for that pollutant. The significant impact criteria are shown in Table 8-3. If the impact is expected to be greater than these criteria for a specific pollutant, baseline monitoring for that pollutant is required. In addition to air quality monitoring, the source is also required to obtain site-specific meteorological monitoring data to be used in the air quality impact analysis (AQIA).

■ **TABLE 8-3** Significance Levels for Air Quality Impacts[a]

Pollutant	Annual	24-Hour	8-Hour	3-Hour	1-Hour
SO_2	1	5	—	25	—
TSP	1	5	—	—	—
PM_{10}	1	5	—	—	—
NO_x	1	—	—	—	—
CO	—	—	500	—	2000
O_3	—	—	—	—	[b]

[a]This table does not apply to Class I areas. If a proposed source is located within 100 kilometers of a Class I area, an impact of 1 $\mu g/m^3$ on a 24-hour basis is significant.

[b]No significant ambient impact concentration has been established. Instead, any net emissions increase of 100 tons per year of VOC subject to PSD would be required to perform an ambient impact analysis.

A total of at least 12 months of on-site data must be obtained unless it can be demonstrated that 4 months are sufficient to adequately characterize the area or that existing off-site data are representative, and the local permitting authority agrees to the reduced on-site monitoring. One important point to note is that at least 90 percent of all possible data in the 12-month period must be obtained and validated in order for the monitoring to be accepted. In remote locations, it is often difficult to obtain that level of data capture. Another key point is that the data obtained must be within the most recent 5 years. This can become a problem for projects that have long lead times due to financial or other planning considerations.

If no sources of emissions have been constructed in the impact zone of the proposed source, the baseline concentrations will simply be the measured concentrations at the time of the preconstruction monitoring program. If, however, there have been sources of pollution constructed in or near the impact zone after the baseline date, then the baseline concentrations will be the measured concentrations minus the impact of each of the new sources built since the baseline date. In this case, a "near source" is defined as any source whose impact zone intersects the impact zone of the proposed source. (Note that this analysis is done on an impact-zone basis, not on an AQCR basis.)

PSD Increments

The PSD program is designed to prevent "significant" deterioration of air quality. It defines significant impacts as impacts greater than PSD increments and/or those impacts that adversely affect soils, vegetation, visibility, or air-quality-related values, as described in later sections.

PSD increments have been established by the 1977 CAA and are shown in Table 8-4. Note that there are three sets of increments for three different classes of areas. Note also that the increments are pollutant- and averaging time-specific.

The three classes of areas are as follows:

Class I. These areas are international parks, national parks over 6000 acres in size, national wilderness areas over 5000 acres in size, national memorial parks over 5000 acres in size, and other areas redesignated Class I by the local government pursuant to the CAA. All Class I areas except the redesignated Class I areas are termed "mandatory Class I" areas. Mandatory Class I areas have a stronger legal status than redesignated Class I areas.

■ TABLE 8-4 PSD Increments[a]

Pollutant	PSD		
	Class I	Class II	Class III
Sulfur oxides (sulfur dioxide)			
Annual	2	20	40
24 hours	5	91	182
3 hours	25	512	700
Particulate (PM_{10})			
Annual	4	17	34
24 hours	8	30	60
Carbon monoxide			
8 hours	—	—	—
1 hour	—	—	—
Ozone			
1 hour	—	—	—
Nitrogen dioxide			
Annual	2.5	25	50
Lead			
Quarterly	—	—	—

[a]Concentrations in $\mu g/m^3$ and arithmetic average [40 CFR 52.21 (C)].

Class I areas deserve the most stringent protection and have the lowest increments. Note that some of these increments are very small. It would not be unusual for a single large source located as far as 50 kilometers from a Class I area to completely consume the Class I increment.

Class II. These are all areas of the United States that are not Class I and have not been designated Class III by the governor of the state.

Class III. These areas are those designated by the governor of a state as requiring less protection than Class II areas. Practically speaking, it is highly unlikely that a governor would designate such an area, and there are no Class III areas to date. Therefore, the entire United States is either Class I or Class II.

Although the CAA allows for the government to redesignate a Class II area as Class I, this option has been rarely exercised, and to date, the only redesignated areas have been a few Native American tribal lands: the Northern Cheyenne, Fort Peck, and Flathead Indian Reservations in Montana and the Spokane Indian Reservation in Washington (EPA, 1990). Therefore, in practice, one can assume that essentially only national parks and major wilderness areas are Class I except for the four tribal lands noted.

PSD increments are *de facto* new, more stringent ambient air quality standards in that, once a baseline concentration is established, the total ambient air quality impact cannot exceed the baseline plus increment. For example, if the baseline concentration for sulfur dioxide is 30 micrograms per cubic meter and the PSD increment is 91 micrograms per cubic meter, the ambient air quality impacts of all sources affecting the baseline area cannot exceed a total of 121 micrograms per cubic meter, even if the NAAQS is 365 micrograms per cubic meter.

On the other hand, if the baseline concentration is 320 micrograms per cubic meter, and the increment is 91 micrograms per cubic meter, the total ambient air quality impact of all sources affecting the area cannot exceed the NAAQS of 365 micrograms per cubic meter. (This is due to the fact that the baseline concentrations plus the increment in this example would be 320 + 91 = 411 micrograms per cubic meter; but the total of baseline plus increment can never exceed the NAAQS.) Obviously, in relatively clean areas, the PSD increments force the area's air quality to remain much better than the NAAQS, which is, in fact, the purpose of the PSD program.

Air Quality Impact Analysis

In order for a proposed new source to obtain a permit, the source must demonstrate that the ambient air quality impact will not cause an exceedance of the NAAQS, state ambient air quality standards (SAAQS), and the PSD increments applicable for the class of area affected by the source (i.e., within the impact circle). This demonstration is accomplished through the air quality impact analysis (AQIA).

The AQIA involves three steps:

1. Perform ambient air dispersion modeling (Chapter 4) to evaluate the impact of the source for each emitted pollutant for the various standard averaging times.
2. Identify all stationary sources (major and minor) that have been built since the baseline date that could affect the same impact circle as the proposed source and determine the effect of those sources on the impact zone of the proposed source.

3. Add the impact of the proposed source plus other sources built since the baseline date to the baseline concentrations and demonstrate that the sum of the impacts is less than the applicable PSD increments, NAAQS, and SAAQS.

If the baseline meteorological and ambient air quality impact monitoring data are current (i.e., conducted in the 12 months immediately prior to proposing the new source), one might think that the monitoring would have already indicated the total contribution of all sources except for the proposed source. (In other words, one might think that monitoring data could substitute for step 2 above.)

Technically speaking, this is the case. However, due to meteorological variability, even 12 full months of monitoring do not adequately characterize the full range of possible impacts of a source. Therefore, the EPA requires the proposed source to explicitly model the potential ambient air quality degradation of existing sources that could affect the same impact zone as the proposed source. In order to limit the number of potential sources that might have to be examined, the EPA has stated that only those sources located within 50 kilometers of the outer boundary of the proposed source impact circle need to be identified and modeled. The 50-kilometer distance was chosen because Gaussian dispersion models are not generally reliable beyond 50 kilometers. (As stated in Chapter 4, even 50 kilometers is a much farther distance than what should be used in dispersion modeling; but it is the distance required by regulations.) Figure 8-1 shows how the impact circle and 50-kilometer distance for evaluating sources interact when evaluating ambient air quality impacts of a proposed source.

Emission Scenarios for the AQIA

Normally, the maximum ambient air quality impact of a source occurs when the source is operating at full capacity, and the emissions are the greatest. This is nearly always the case for annual average impacts. However, there are other situations where the ambient air quality impacts occur at less than full capacity.

For example, assume that a source has a relatively low stack height. Further assume that the source operates at 50 percent of capacity (and, therefore, 50 percent of maximum emissions), and the stack exit velocity is less than 50 percent (e.g., 30 percent of full capacity). In this case, the low stack exit veloc-

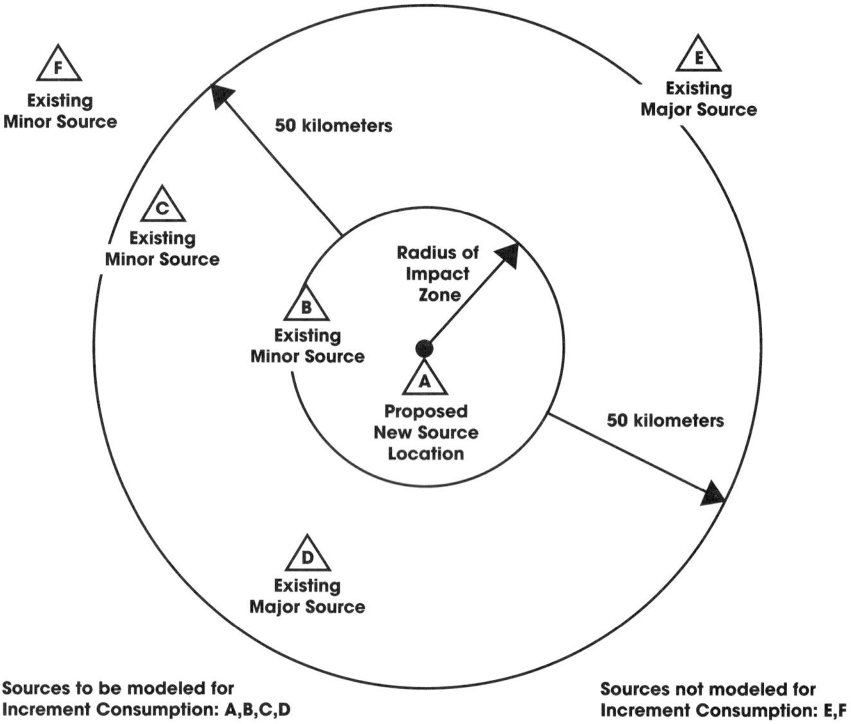

■ **FIGURE 8-1 Selecting Sources for PSD Increment Consumption Evaluation.** *Source:* After USEPA (1990).

ity could cause a lower plume rise and higher near-source impacts than when the source was operating at full capacity and emissions. This is due to the fact that the ambient dispersion models are not linear with respect to parameters other than emission rate (Chapter 4).

Therefore, the AQIA must demonstrate that all operating conditions have been examined, and the worst-case condition leading to the greatest ambient impact has been identified and analyzed.

Good Engineering Practice Stack Height

In the 1960s, some large sources were built with very tall stack heights (i.e., stack heights greater than what would normally be installed at a plant). The purpose of the tall stacks was to disperse the pollutants before they reached the ground and minimize the ambient air quality impact of the source. During the debate over the 1977 CAA Amendments, this practice was soundly criticized. Therefore, the 1977 CAA prohibits a source from using dispersion techniques [such as merged gas streams, intermittent controls, or stack heights greater than that required by good engineering practice (GEP)] to reduce ambient air quality impacts. Thus was born the GEP stack height issue and the GEP stack height rule.

The EPA has defined GEP stack height [40 CFR 51.100(ii)] to be the larger of the following:

1. 65 meters
2. Height equal to the height of nearby structures plus 1.5 times the height or projected horizontal dimension, whichever is smaller (termed the H + 1.5L rule)
3. Height demonstrated by fluid modeling or field study as being necessary to avoid excessive concentrations of pollutants near the stack caused by atmospheric downwash, wakes, or eddy effects

A proposed source can build as tall a stack as it wants to, but the AQIA cannot use in the model a height greater than the GEP height. (This is termed "not getting credit for stack heights greater than GEP.") Normally a source will build the stack at GEP height.

Visibility, Soils, and Vegetation Impacts

Thus far, the AQIA discussion has focused on the criteria pollutants and direct impacts of those pollutants as measured by comparison of the impacts to PSD increments and ambient air quality standards.

The PSD program also requires, however, an examination of potential indirect air quality impacts, such as adversely impairing soils, vegetative growth, or visibility. These indirect impacts are those that could occur even if the proposed source does not cause an exceedance of the increments or ambient air quality standards. There are no rigorous methods for assessing these types of impacts; however, the EPA has promulgated a dispersion model that supposedly predicts visibility impacts of proposed sources (Chapter 4, Table 4-7), and the U.S. Forest Service (USFS) and National Park Service have promulgated guidance that provides a methodology to assess soils and vegetative impacts (e.g., USDA, 1992).

One key point to remember, however, is that, by legislation, only vegetation deemed to have commercial value needs to be analyzed. (This is in contrast to the case where vegetation impacts are analyzed as part of the Class I AQRV analysis. In the Class I impact situation, *all* vegetation needs to be analyzed.)

Air Quality Related Values Impacts Analysis

For Class I areas (only) the impact of the proposed source on air quality related values (AQRVs) is required. AQRVs are defined as including, but not limited to, impacts on soils, vegetation, and visibility. Because AQRVs are not limited, the Federal Land Managers (FLMs) have broadly interpreted this requirement to include evaluating all conceivable potential air quality impacts. The FLM is the agency responsible for the Class I area, which is usually the National Park Service (NPS), but can include the USFS or the U.S. Fish & Wildlife Service (USFWS). For redesignated Class I areas, the governing agency for the redesignated area is the FLM (e.g., the specific Native American tribe and/or the Bureau of Indian Affairs for Tribal Lands). The Bureau of Land Management (BLM) is also often involved as an FLM.

Under the 1977 CAA, the FLM has an affirmative responsibiltiy to protect AQRVs in Class I areas. AQRVs include such issues as

1. Acid deposition
2. Visibility (contrast, color, visual range, regional haze, plume blight, scenic beauty, etc.)
3. Cultural, archaeological, and paleontological preservation
4. Ecosystem impacts on flora or fauna, including growth, mortality, reproduction, diversity, visible injury, succession, productivity, and abundance
5. Water quality, including metals concentration, pH, dissolved oxygen, anion and cation concentrations
6. Soils impacts, including cation exchange capacity, metals deposition, pH
7. Any other potential impact that could be considered adverse to the Class I area

The FLMs have established various criteria for determining adverse impacts, but these criteria are in themselves generally stated. For example, the NPS considers impacts to be adverse if they

1. Diminish the area's national significance
2. Impair the structure or functioning of ecosystems
3. Impair the quality of the visitor experience

Accordingly, one must look very broadly and address a wide range of possible impacts when conducting the AQRV analysis. Consultation with the FLM, review of FLM guidance, and review of previously issued PSD permits and applications are useful references to help focus the AQRV analysis.

One of the issues related to AQRVs is visibility protection in national parks. The problem is how to define what is "in" a national park. Literally, one would think that "in" a park means a person inside the park viewing a vista also in the park. However, because many of our national parks are in generally picturesque settings, many would argue that visibility protection in national parks includes views from within the park looking out and views of persons outside the park looking at the scenery within the park. These types of views are termed "integral vistas," in that the views are integral to the enjoyment of the park.

In the 1980s there was considerable debate over this issue, and the EPA left the decision up to the FLMs to identify and then regulate integral vistas. Although a draft list of integral vistas was developed and published by the FLMs, only one park in Maine actually formally designated integral vistas. As far as the EPA is concerned, then, the only integral vistas explicitly requiring protection are those in Maine. Practically speaking, however, through the public review and comment process as well as regulations that may be promulgated by the local permitting authority, integral vistas are and will be protected as part of the AQRV review, even if those vistas were not formally promulgated by the FLM. Accordingly, one must look closely at all of the Class I areas even remotely associated with the location of a proposed source to ensure that all possible issues are identified and addressed. If not, the permit application could be significantly held up or denied. Early involvement and consultation with the FLM is very important to an efficient permitting process. The EPA gives great deference to FLM comments and concerns.

Best Available Control Technology (BACT) Demonstration

To some, one of the more frustrating aspects of the PSD program is that, regardless of the ambient air quality impact of the proposed source, the source must install Best Available Control Technology (BACT). In other words, even if without controls or with controls that are not as stringent as BACT, the source ambient air quality impacts (including AQRVs) are acceptable, the source must still install the most stringent controls that are technically and economically feasible.

The BACT concept is used throughout the air quality arena. For example, in the Los Angeles area, BACT is required to be installed on all sources, regardless of the magnitude of emissions from that source. Therefore, the full definition of BACT, as stated in the CAA Section 165(a)(4), will be quoted herein. BACT is defined as

> an emission limitation (including visible emission standard) based on the *maximum degree of reduction* for each pollutant subject to regulation under the Clean Air Act which would be emitted from any proposed major stationary source or major modification which the Administrator *on a case-by-case basis,* taking into account energy, environmental, and *economic impacts* and other costs, determines *is achievable* for such source or modification through application of production processes or available methods, systems, and techniques, including fuel cleaning or treatment or innovative fuel combustion techniques for control of such pollutant. [Emphasis added.]

The definition goes on to state that

> If the Administrator determines that technological or economic limitations on the application of measurement methodology to a particular emissions unit would make the imposition of an emissions standard infeasible, a design, equipment, work practice, operational standard, or combination thereof, may be prescribed instead to satisfy the requirement for the application of BACT. Such standard shall, to the maximum degree possible, set forth the emissions reduction achievable by implementation of such design, equipment, work practice or operation, and shall provide for compliance by means which achieve equivalent results.

In practice, the key elements of the BACT definition are as follows:

1. It must provide the maximum degree of reduction (i.e., be the most stringent).
2. It is determined on a case-by-case basis.
3. Economic impacts must be considered.
4. It must be technically achievable.
5. It can consist of work practice and other operational limitations as well as, or instead of, control equipment.
6. The degree of reduction must be measurable, or at least provide for compliance with equivalent results to a measured emissions reduction.

Case-by-Case Basis One of the key things to note about BACT is that it is a case-by-case determination made by the local permitting authority with EPA oversight. Therefore, BACT will consistently become more stringent (see the next section on "top down" BACT determinations) with time. What this also means is that an equipment vendor cannot guarantee that a particular control device will meet BACT. All the vendor can guarantee is that the last time the device was permitted, a certain permitting authority accepted the technology as BACT. This causes uncertainty in equipment design and difficulty in planning new sources.

There are, however, at least two available guides that help in determining BACT for a particular type of source. The most commonly accessed and widely available is the RACT/BACT/LAER (RBL) Clearinghouse maintained by the EPA. The Clearinghouse publishes, in both hard copy and electronic versions, a list of all permitted new sources and the degree of control required for each of those sources. One can access the Clearinghouse and determine the most stringent permit requirements that have been issued to date for a particular type of source category. This degree of control will be at least required on the next proposed source in that category. In other words, no new source can be permitted with emissions greater than the most stringent previously permitted source.

The second source of information is the local permitting authority's database of permits and BACT determinations. Various agencies have these data available in various formats and accessibility. One source that is often used is the South Coast Air Quality Management District *BACT Guidance Manual* (SCAQMD, 1988). This manual is convenient because it summarizes in tabular format BACT determinations made by the SCAQMD for a wide variety of sources. Because the Los Angeles area has the worst air quality in the nation, it is generally considered that if a device meets BACT in SCAQMD, it will likely meet BACT in any other permitting jurisdiction. The BACT guidance manual is available for purchase from the SCAQMD.

The case-by-case nature of BACT is different from all but one of the other federally mandated control technology requirements. MACT and NESHAPS for air toxics, as well as NSPS for new sources, are all generic to a source category or type of process and are the same nationwide. Reasonably Available Control Technology (RACT) in nonattainment areas (Chapter 9) is also generic to source categories in a local jurisdiction. The one other exception to the generic source category type of control programs is the other nonattainment program, Lowest Achievable Emission Rate (LAER), which is similar to BACT in that it is a case-by-case determination.

Technically and Economically Feasible The uncertainty related to case-by-case determinations is further exacerbated when the concept of technical feasibility enters the determination. The old adage about "give an engineer enough time and money and he or she can accomplish anything" is problematic when stating that the control systems must be only technically feasible in order to be required.

"Technically feasible" means that the proposed source must install the most stringent control system that could be installed on the source. This is relatively easy to determine if another source that is similar to the proposed source has already installed a certain type of control. This is termed "demonstrated in practice." Obviously, if a certain type of control system has been demonstrated in practice, a system at least as stringent is required to be installed.

On the other hand, if a more stringent system has not yet been installed on a certain proposed source type, but engineering judgment indicates that it would probably work, then it is technically feasible and must be installed. Engineering judgment is subject to much discussion. Therefore, the EPA has established a "top down" BACT determination policy. This policy requires the proposed source to do the following:

1. Identify all possible control systems for the proposed source, whether or not the systems have been achieved in practice.
2. Evaluate each control system on the list and determine if it is technically feasible. Even if the system has not yet been installed, if engineering judgment indicates that it is *possible* to be installed, regardless of cost, then it must be considered for the next step.
3. Rank the remaining control systems in decreasing order of stringency.
4. Evaluate the energy, environmental, and economic costs of each control system. If the top system (most stringent) is considered too costly, then proceed to the next most stringent.
5. Select the most stringent option not rejected.

In order to determine what is too costly, the energy, environmental, and economic costs as well as the energy and environmental benefits must be estimated for each technology. Usually, the question boils down to economic costs, although the energy and environmental considerations cannot be ignored. When considering costs, capital, operations, maintenance, and reliability costs are included in the evaluation.

What is "too costly" is also a judgment call. Usually, the regulatory agency starts with the position that, no matter how great the cost, it is economically feasible, whereas the proposed source considers any additional cost as being economically infeasible. One of the methods for reaching consensus on this issue is to plot costs versus control stringency in order to summarize the cost analysis for each technically feasible control system. [For example, plot costs on the y-axis and increasing stringency (i.e., percent removal of pollutant) on the x-axis.] When this is done, it is often found that there is a break point at which costs increase exponentially for a minimal increase in stringency. The most stringent technology just prior to that break point is considered as economically feasible. Figure 8-2 exemplifies this concept.

Another method used to evaluate costs is for the local permitting authority to publish a cost-effectiveness number. For example, in the Los Angeles area the local permitting authority, SCAQMD, has stated that control technology that costs more than $17,500 per ton of pollution removed (in 1987 dollars) is "too costly." For example, if a source was permitting a new paint booth that used 10 gallons of coating per day, 260 days per year, with a VOC content of 6 pounds per gallon of coating, the uncontrolled emissions would be $6 \times 260 \times 10 = 15{,}600$ pounds of VOC per year. If a control system

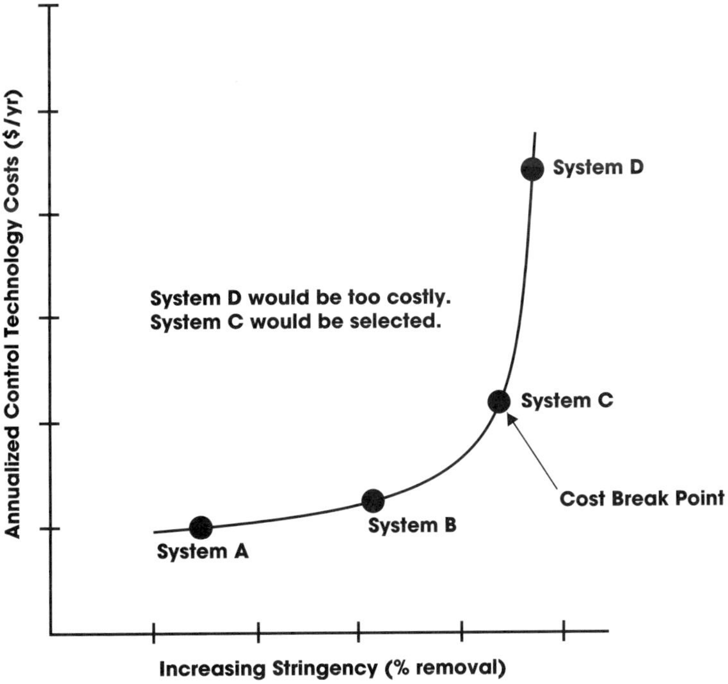

■ **FIGURE 8-2** Evaluating BACT cost-effectiveness. *Source:* After USEPA (1990).

(e.g., activated carbon) removes 95 percent of the emissions, the amount of pollution removed would be 0.95 × 15,600 = 14,820 pounds per year, or 7.41 tons per year. The SCAQMD would consider any control system costing more than 7.41 × \$17,500 = \$129,750 too costly.

Compliance Monitoring After a BACT system is selected, the permit requirement to install the system must be in terms of control efficiency—for example, percent removal and/or mass emission rate. One must be careful when specifying emission rates or percent removal requirements to also specify at what conditions those removal efficiencies occur. For example, it is not unusual for a control system to be more effective, from a percent-removal standpoint, the greater the inlet concentration of the pollutant, and much less efficient, from a percent-removal standpoint, when the inlet pollutant loading is low. Therefore, BACT requirements should be stated similar to the following example:

> No more than 50 pounds per day of VOC emissions from the control system and at least 98 percent removal of VOCs when the inlet rate is greater than 1000 pounds per day and at least 95 percent removal when the inlet rate is less than 1000 pounds per day. The maximum allowed inlet VOC rate is 2500 pounds per day.

In this example, when the inlet pollutant rate is 1000 pounds or less, the emissions will be 5 percent of the inlet rate—that is, 50 pounds or less. When the inlet rate is between 1000 and 2500 pounds per day, the emissions will be 2 percent of the inlet, or between 20 and 50 pounds per day.

Another key point of this example is that emission rates must be stated in terms of an averaging time. All too often, equipment vendors quote removal efficiencies or emission limits but do not specify the time averaging in the quote. This can lead to serious compliance problems. For example, if the proposed source accepts a permit condition of no more than 40 parts per million of nitrogen oxides in the equipment exhaust (a common emission limit) without specifying an averaging time, the limit will be considered to apply at all possible times, even as short as a 5-minute average. However, control equipment efficiency varies, and some sort of averaging time must be stated. For example, equipment vendors usually consider the 40-parts-per-million emission limit to apply over a 1-hour averaging time. Obviously, if this situation is not clarified, there could be compliance problems.

Public Participation

One of the problems with PSD New Source Review is the amount of time it takes to get the permit. This is due to the complexity of the permit review process as well as the fact that the EPA cannot issue the permit without allowing for involvement of the public (through public notice and public hearings), the FLM of nearby Class I areas (all Class I areas within 100 kilometers), and nearby states (usually those states within 50 kilometers) that could be affected by the proposed source. In addition, the EPA and local permitting authorities usually also allow significant involvement of the FLMs of nearby (usually 50 kilometers) Class II areas if the areas are national forests, national monuments, or other lands that are generally considered by the public to deserve special protection (even though the regulations and legislation do not explicitly require such involvement).

Overall PSD Application Process The PSD NSR process is as follows. For each step, an approximate time to accomplish the step is also given:

1. Hold a presubmittal meeting with the EPA and/or local permitting authority to discuss the project and identify who should or will be involved in reviewing or providing input to the permitting authority (starts the process, week 0).
2. Prepare a presubmittal work plan (usually including a monitoring and modeling plan) and submit it to the agency and other interested parties for review and approval (4 weeks).
3. Complete 12 months of baseline monitoring (total of 68 weeks: 8 weeks to install the monitoring equipment, 52 weeks of monitoring, and 8 weeks after completion of the monitoring to report the data).
4. Submit the permit application, including modeling and monitoring results and BACT demonstration (16 weeks).
5. The agency reviews the permit application, requests additional information, and then finally makes a determination that the application is complete (this process can take more than a year, but the typical time is about 16 weeks).
6. The agency reviews the permit application and makes an initial determination of approval (usually requires 52 weeks, but can, by regulation, take 18 months).
7. The agency issues a public notice of its determination to issue the permit and holds a public hearing on the determination (12 weeks).

8. After incorporating or addressing public comments, a permit is issued (8 weeks).

As can be seen from the above, the total process can take 3 years to complete. Obviously, if existing baseline data are available, and the permit application and process are straightforward, the total time can be reduced to 6 to 9 months. However, it should be assumed that NSR permitting of a major source is a very time-consuming process.

Role of Public and FLM The CAA requires the permitting authority to solicit and address issues raised by the public, FLM, and nearby states prior to issuing or denying a permit application. However, the Act does not require the EPA to take the specific action requested by the public, FLM, or states. Even though the FLM has, by law, an affirmative responsibility to protect air quality in their Class I areas if, after considering the facts, the EPA believes that the permit can be issued and the source built without adversely affecting air quality, despite the objections of the FLM, the permit can be issued. In practice, however, this rarely occurs, and if the FLM opposes a project, the project is rarely built. Therefore, it is incumbent on the proposed source to solicit and address FLM comments.

EPA DELEGATION OF PSD AUTHORITY TO LOCAL JURISDICTIONS

Throughout this discussion of the NSR process, the terms "local permitting authority" and EPA were used interchangeably. The local jurisdiction (e.g., state, county, or district air quality agency) will normally have the lead responsibility to review, approve, and issue the permit for a major new source. However, this can only occur if the EPA has delegated its responsibility to that local agency. If not, the local EPA region will review, approve, and issue the permit. One thing to keep in mind, however, is that in the case of major sources, EPA oversight is often quite stringent, and it is often the case that although a local agency has permitting authority for new sources, it will actively seek and respond to EPA input. Therefore, the environmental manager must not get a false sense of security if he or she has only been dealing with a local agency during the presubmittal process for a major source.

VISIBILITY TRANSPORT COMMISSION

One of the new aspects of the 1990 CAA is establishment of the Grand Canyon Visibility Transport Commission. The purpose of this Commission is to evaluate visibility impacts in the Grand Canyon National Park and recommend programs to the EPA to remedy existing and prevent future visibility degradation in the park. The Commission has been deliberating the issues since 1992 and will have its recommendations to the EPA in early 1996. Although, as previously discussed, the PSD program protects visibility in the park by requiring an analysis of new major sources, the Visibility Transport Commission can recommend regulations for new minor sources and all existing sources as well.

One of the more serious options under consideration by the Commission is to establish "clean air corridors," where all new sources would require nonattainment-style external offsets prior to being issued a permit. In addition, just as in the nonattainment areas, all existing sources in the corridors would be required to install Reasonably Achievable Control Technology (RACT). This new area of regulation could result in the application of nonattainment stringency to attainment areas in order to protect the esthetic visibility of the Grand Canyon.

NEW SOURCE REVIEW PERMIT STREAMLINING

There are a number of difficulties with the current NSR regulatory system. These difficulties have been pointed out by industry, regulatory agencies, and environmental groups. In 1992 the EPA began a detailed review of the NSR regulatory system, and in July 1994 it proposed a revised system in draft discussion form. The proposed revisions were designed to streamline NSR permitting. Among the issues being evaluated are the following:

1. How to evaluate, account for, and remedy Class I area impacts caused by regional, distant, and/or local minor sources.
2. How to evaluate potential ozone impacts since ozone does not have an increment and is an indirect result of ozone precursors (NO_x and VOC) and not a directly controlled pollutant.
3. Establishment of an emissions offset requirement for Class I areas already impacted.

4. Expansion of the preconstruction monitoring requirement to include monitoring in Class I areas as well as postconstruction monitoring in and near Class I areas.
5. Streamline BACT determinations by
   a. Establishing a technology cutoff date
   b. Publishing cost–benefit data in the RBL Clearinghouse
   c. Publishing guidance on how to consider foreign technology not yet used in the United States
   d. Developing presumptive BACT for different types of sources
   e. Implementing a fast track BACT determination system
6. Implementation of a Plantwide Applicability Limit (PAL), similar to a plantwide emissions cap, and allow facilities to add emissions from new devices at the facility without triggering NSR, as long as the PAL was not exceeded.

CHAPTER 9

NONATTAINMENT

If a region does not meet the NAAQS, the area is termed a *nonattainment* (NA) area. New Source Review in nonattainment areas proceeds similarly to NSR in attainment areas (i.e., PSD and minor source permitting, Chapter 8). However, the thresholds are lower, and the requirements are more stringent. This chapter discusses how NA areas are designated, impacts of nonattainment designation on existing sources, and NSR in NA areas. The fundamental purpose of the various nonattainment area regulatory programs is to bring the area into attainment. This means to reduce the level of pollution in the area to levels below the NAAQS, which, by definition, are the levels necessary to protect the public health and welfare with an adequate margin of safety. Nonattainment and acid deposition are the two main issues that eventually brought enough political urgency to get the 1990 CAA amendments enacted.

NONATTAINMENT AREA DESIGNATIONS

Nonattainment Areas

In general, NA designation occurs by air quality control region (AQCR). AQCRs are, in general, specified by political boundaries (e.g., county lines) and not by air quality related criteria (e.g., coastal locations versus inland locations). Recognizing that meteorological and topographical differences within an AQCR can greatly affect the air quality, there are often subsets of NA designations within an AQCR.

These subset designations can be by "air basins" or simply by artificial political distinctions (e.g., township and range designations). An air basin is an area that has similar meteorological and topographical characteristics such that the air quality in the area is somewhat similar. Two examples of these types of distinctions follow.

In San Diego County, California, the eastern half of the county is attainment for ozone, while the western half is nonattainment. This difference is related to meteorology and topography, but most significantly to population distribution. The dividing line is a series of township and range designations incorporated into the regulation. On the other hand, in Los Angeles County, the Desert Air Basin is separated from the Coastal Air Basin, and the dividing line is the mountain range immediately east of the city that separates the desert climate from the coastal climate.

In addition to allowing for subsets of AQCRs to be designated attainment or nonattainment, the 1990 CAA extends the boundary of serious CO or ozone nonattainment areas to include the Metropolitan Statistical Area (MSA) or Consolidated Metropolitan Statistical Areas (CMSAs). This is due to the fact that much of the urban pollutant loadings are related to vehicular travel, and many persons who work in cities live outside of the city. This extension to MSAs or CMSAs occurs regardless of the AQCR boundary.

Appendix L contains a list of current NA designations. Note that attainment and nonattainment designations are pollutant-specific, but not averaging time-specific. (In other words, if an area meets the 24-hour SO_2 NAAQS but does not meet the annual NAAQS, the area is designated nonattainment even though the 24-hour NAAQS is met.)

Basis for Nonattainment Designations

In general, an NA designation is made based on actual monitoring data, rather than predictive modeling. However, SIP programs to bring an area into attainment are evaluated on a predictive modeling basis. The monitored data considered for designation are those from the immediately previous 8 quarters (2 years). Even if only a single monitoring station in an area monitors an exceedance of the NAAQS, if that monitor was not affected by obviously erroneous or nonrepresentative conditions, then the area is nonattainment.

In addition to comparing the "worst monitor" to the NAAQS to determine attainment, the 1990 CAA established a new concept of "design value" for classifying the severity of the area's NA status. The design value for ozone is the fourth highest value observed over a 3-year period of monitoring in the area. The design value for CO is the second-highest nonoverlapping 8-hour concentration recorded in an area over a 1-year period.

Nonattainment Area Classification Status

Carbon Monoxide and Ozone Table 9-1 shows the 1990 CAA classification scheme for ozone and CO nonattainment areas. As can be seen, ozone areas are classified from marginal to extreme; however, there is only one extreme ozone NA area currently in the United States: the Los Angeles area.

Table 9-1 also shows the attainment deadlines as required by the CAA. As noted, the worse the area is, the longer time it has to come into attainment. However, as will be seen shortly, worse areas have much more stringent interim requirements to reach attainment.

Particulate Matter The 1990 CAA initially classifies all PM_{10} NA areas as "moderate" with an attainment deadline of November 15, 1994. If, however, the state or the EPA determines that the area cannot come into attainment by 1994, the area will be reclassified as "serious," and the attainment date moved to 2001.

Nitrogen Oxides For NO_x, there are no different NA classifications. However, NO_x emissions are affected by the ozone NA classification status in

■ TABLE 9-1 Classification Design Values and Attainment Deadlines for Nonattainment Areas

Ozone[a]			CO[a]			PM_{10}[b]	
Status	Design Value (ppm)	Attainment Deadline	Status	Design Value (ppm)	Attainment Deadline	Status	Attainment Deadline
Marginal	0.121–0.137	Nov 15, 1993	Moderate	9.1–16.4	Dec 31, 1995	Moderate	Nov 15, 1994
Moderate	0.138–0.159	Nov 15, 1996	Serious	16.5 or more	Dec 31, 2000	Serious	Nov 15, 2000
Serious	0.160–0.179	Nov 15, 1999					
Severe-15	0.180–0.189	Nov 15, 2005					
Severe-17	0.190–0.279	Nov 15, 2007					
Extreme	0.280 or more	Nov 15, 2010					

[a]Two 1-year extensions are available for CO and ozone.
[b]An extension to 2001 is available for PM_{10}.

that NO_x emissions also affect ozone formation. In general, NO_x emissions add to ozone formation. However, it is now recognized that in certain situations, reducing NO_x emissions can actually cause an *increase* in ozone. This is because, depending on the ratio of VOCs to NO_x in the atmosphere, NO_x can scavenge ozone in the atmosphere. Therefore, how NO_x is treated depends on the specific situation of the local AQCR and/or NA area.

Sulfur Oxides Just as for NO_x, there are no separate classifications for SO_2 nonattainment. The number of SO_2 NA areas is relatively small. The primary issue associated with SO_2 is acid deposition, which is addressed in Title IV of the 1990 CAA (Chapter 11).

Lead One of the most successful clean air programs to date is the phaseout of lead in motor fuels. This program has resulted in nearly all areas of the United States being in attainment for lead.

NONATTAINMENT AREA REDESIGNATION

Designation and redesignation of nonattainment or attainment areas is done by the EPA. However, the EPA will, of its own accord (based on the monitored data), designate an NA area; but redesignation to attainment requires a filing and demonstration of attainment by the state. Eight full quarters of monitoring indicating attainment of the NAAQS is but one of several requirements that must be met before redesignation. The requirements are as follows:

1. The area under consideration must not be contributing to nonattainment in nearby areas.
2. There must be a SIP submittal indicating that the area will remain in attainment for at least 10 years. The plan must include an 8-year revision trigger demonstrating that attainment will still occur at the 10-year interval, and the plan must have contingency provisions demonstrating that monitoring of exceedances will trigger the previous nonattainment controls.
3. The improvement must have been due to permanent and enforceable conditions. For example, reaching attainment because of an economic recession will not be sufficient.

Appendix M contains a memorandum from the EPA that describes how the Agency handles redesignation requests.

REQUIRED GENERALLY APPLICABLE NONATTAINMENT PROGRAMS

General Requirements to Reach Attainment

For all nonattainment areas, states are required to incorporate into their State Implementation Plans (SIPs) various programs to ensure that the areas reach attainment by the deadlines. As the result of the 1990 CAA, states were required to resubmit their SIPs by November 15, 1992, to incorporate the new CAA requirements. Among the new SIP requirements per the 1990 CAA are

1. Contingency plans for emissions presenting imminent and substantial danger
2. Technical assistance program for small businesses
3. Annual increment reduction requirements
4. Creation of detailed current emission inventories

Reaching attainment, except for lead, has been an ephemeral goal. Attainment was first legislated by the 1970 CAA with a target date of 1975. However, attainment was not achieved, and the 1970 Act essentially called for a ban on new construction in the NA areas. The EPA demurred, however, and the 1977 CAA Amendments established a new attainment date of December 31, 1982 for all pollutants except CO and ozone, with December 31, 1987 as the target date for CO and ozone. However, that date was also missed, and the 1990 CAA Amendments established a series of later NA dates, as indicated previously.

The current nonattainment area regulatory requirements are a combination of requirements in place before the 1990 CAA and the new 1990 CAA requirements. These requirements can be summarized as follows:

1. A demonstration that Reasonable Further Progress (RFP) will be made and that all Reasonably Available Control Measures (RACM) have been required in the NA area
2. Reasonably Available Control Technology (RACT) for existing sources
3. Lowest Achievable Emission Rate (LAER) for new sources
4. Reduced thresholds for major sources in nonattainment areas, depending on severity classification
5. Emissions offsets for new major sources at various offset ratios, depending on severity classification
6. Demonstration that there will be a net air quality benefit when a new source or modification is permitted
7. Highway plan conformity review

Reasonable Further Progress and Reasonably Available Control Measures (RFP and RACM)

The CAA requires SIPs to demonstrate to EPA how the area will reach attainment. At a minimum, the SIPs must include all measures considered by the EPA to be RACM. However, the states can and must incorporate more stringent measures as needed to reach attainment.

Furthermore, the 1990 CAA requires SIPs for all ozone nonattainment areas to achieve a 15 percent reduction in VOC emissions by November 15, 1996. All severe, serious, or extreme ozone nonattainment areas must achieve a 3-percent-per-year annual reduction in VOC emissions unless the state uses all measures required by the next higher classification and achieves all measures within 6 years. In that case, the 3 percent annual reduction can be waived by the EPA.

Prior to implementing the RACM measures needed to meet RFP, the local permitting authority publishes its “game plan” in a document termed the Air Quality Management Plan (AQMP). The AQMP specifies what measures will be required, by when, and to what degree. It also describes how the emissions inventory will change when the measures are implemented.

From the perspective of an environmental manager, the SIP and AQMP requirements do not directly affect his or her facility until the SIP measures are promulgated as local regulations and rules. However, it is useful to carefully track development of the AQMP and other measures the local permitting authority is considering, because early input at this stage of the debate is often much more effective in helping to achieve a reasonable rule than participation later in the rule development stage.

Reasonably Available Control Technology (RACT)

The SIP must require installation of reasonably available control technology on all existing sources as part of the attainment plan. RACT is less stringent than BACT required of new sources and is the level of control that is demonstrated in practice and is cost-effective. RACT can include work practice and other operational changes that reduce emissions, as well as traditional after-process control technology.

Practically speaking, what is or is not RACT becomes an issue for the environmental manager only when debating a proposed rule, and not as a practical daily compliance issue. This is because RACT is incorporated by the local permitting authority into a series of device-specific rules.

Lowest Achievable Emission Rate (LAER)

LAER is required of all new major sources, in addition to the emissions offset requirement. LAER is similar to the requirement for BACT in attainment areas, except that LAER does not include a cost–benefit analysis. In other words, for a technology to be considered LAER, all that is necessary to demonstrate is that it is technically feasible.

Theoretically, LAER is more stringent than BACT. However, in practice, BACT often defaults to technical feasibility as well (i.e., the local permitting authorities rarely conclude that a particular stringent technology is "too expensive"). Therefore, there is often little practical difference between LAER and BACT. In fact, some local jurisdictions simply equate by regulation BACT and LAER. The RACT/BACT/LAER Clearinghouses (Chapter 8) are a good source of information to determine what is LAER.

MAJOR SOURCE THRESHOLDS

New sources in nonattainment areas are defined just as they are in the attainment NSR program (i.e., PSD, Chapter 8). Just as in the PSD program, new sources include new and modified sources where the modification increases emissions by more than *de minimis* amounts. However, the major source thresholds are lower in NA areas than in attainment areas.

As previously noted, Tables 8-1 and 8-2 show the federal major source thresholds for various nonattainment classifications and pollutants for new and modified sources. Just as in the PSD program, internal offsetting (netting) is allowed in determining whether or not a modification is major. Note that some local permitting authorities use a zero threshold major source definition for nonattainment areas as well.

EMISSIONS OFFSETS

Because nonattainment areas currently have worse air quality than desired, before any new major source of emissions can be permitted in an area, there must be a concomitant reduction in emissions in the area from other sources to more than offset the new source increase. Emissions offsets occur through a requirement that a new source obtain and surrender emission reduction credits (ERCs).

Offset Ratio

The amount of ERCs necessary is at least equal to the amount of proposed emissions for the new source, but also must be somewhat greater than the proposed emissions because of the offset ratio requirement.

For example, assume that an industry is proposing a new source that will emit, under full capacity and maximum emission conditions, 300 tons of VOCs per year in an ozone nonattainment area. If the offset ratio is 1.2 to 1, the proposed source must obtain and surrender 360 tons per year of ERCs (i.e., 1.2×300 tons per year). The offset ratios vary depending on NA severity and are shown in Table 9-2.

Obtaining ERCs

One of the more difficult issues associated with NA permitting is obtaining ERCs. In order for an ERC to be available to offset new emissions, the ERC must be

1. Real
2. Quantifiable
3. Permanent
4. Surplus

The "real" and "quantifiable" terms have the same meaning as in the PSD netting program. "Real" means that only actual emissions within the last 2 years (or other 2-year period within the last 5 years if the most recent 2-year period can be shown to be not representative) are creditable for ERCs. "Quantifiable" means that adequate records and other information must be available in order to demonstrate the actual magnitude of the emissions.

■ **TABLE 9-2** Emissions Offset Ratios[a]

Status	Ratio
Marginal	1.1 to 1
Moderate	1.15 to 1
Serious	1.2 to 1
Severe-15	1.3 to 1
Severe-17	1.3 to 1
Extreme	1.5 to 1

[a]If attainment or SIPs are not reached or submitted by the deadlines, the offset ratio goes to 2 to 1.

The other two terms, "permanent" and "surplus," have slightly different meanings in NA than in PSD. Permanent means that the emissions reductions must be federally enforceable. The SIP is a federally enforceable document. Therefore, if a reduction is taken and a new permit is filed with the reduced emission limit in place, the reduction would be considered permanent. If, however, there is not a SIP applicable to the reduction (e.g., the device is not required to be permitted), then the reductions would *not* be federally enforceable. In practice, therefore, only those reductions that occur with *permitted* equipment (and the permit is modified to reflect the reductions) can be considered as potential ERCs.

The last term, "surplus," is usually the most troublesome. Emission reductions that occur because of other regulatory programs cannot be considered for ERCs. For example, if the local permitting authority promulgates a rule that requires a reduction on a certain type of source, the required reduction would not count. This is because the local authority has already counted those reductions as helping to bring the area into attainment. When one considers that the local authorities must promulgate rules that require LAER, RACT, and RACM and demonstrate reasonable further progress (RFP), it is not likely that a source has many other options to reduce emissions other than complete shutdown. Therefore, in practice, essentially only equipment shutdowns are available for ERCs.

Another difficulty in determining what emissions are "surplus" is that some local permitting authorities incorporate ERC discounting into their local rules. One example is to allow ERCs only for those emissions from a device that would have occurred if the source had installed BACT (or LAER). This system requires one to assess emissions as if hypothetical control systems were installed on the emitting device. Practically speaking, in this system, usually only 5 percent or less (assuming BACT is usually at least 95 percent control) of the actual emissions of uncontrolled devices would be allowed for ERCs.

The second example is a local agency discounting ERCs by the amount of emissions that would be increased at some other location in the AQCR by some other facility. For example, if there are two industries in the AQCR that make hot tubs and one shuts down, it would be assumed that the other hot tub facility would increase production to compensate for the first facility shutdown. In this case, none of the emissions from the first facility would be allowed as ERCs.

Example ERC Calculation

As an example of how the above concepts of real, quantifiable, permanent, and surplus interact, consider the following example:

> ABC, Inc. is a large facility that currently (in 1994) emits 481 tons per year of VOCs, facilitywide. All of the actual emissions are permitted. The facility wants to install a new process line that will emit, after installation of LAER controls, an additional 70 tons per year of VOCs.
>
> The facility has a long-term history in its current location, and prior to 1986 it emitted 585 tons per year of VOCs. However, in 1986 one of the permitted devices at the facility was removed, decreasing facility emissions by 10 tons per year. Upon removal, the permit for that device was surrendered to the local permitting authority.
>
> In 1990 the facility shut down its permitted "XYZ No. 1" process lines that emitted 60 tons per year of VOCs, thus reducing the emissions for the facility to 515 tons per year. The reason for the shutdown was that the local permitting authority promulgated an emissions reduction rule in 1989 that required XYZ process lines to reduce emissions by 90 percent. The emission reduction rule that was promulgated was included in the state's SIP to bring the area into attainment. The permits for the shutdown process line were surrendered.
>
> In 1991 the facility shut down an unpermitted device (a degreaser) that emitted 4 tons per year of VOC. The degreaser did not require a permit, because it was too small.
>
> In 1992 the facility shut down one of its large degreasers ("Large Degreaser No. 1") that emitted 15 tons per year. The degreaser was permitted, and after shutdown the permit was surrendered.
>
> In 1993, the facility shut down a second large degreaser ("Large Degreaser No. 2") that also emitted 15 tons per year. That large degreaser was permitted because of its size. However, the facility manager was not sure that he might not use the degreaser in the future, so he did not surrender the permit for the degreaser.
>
> The question is, How many ERCs must the facility obtain, if any, prior to construction of the new process line?

The principles of real and quantifiable emissions are established in this example by the fact that the facility maintained records of all the processes and their emissions. Therefore, there was no question by the local permitting authority regarding the magnitude of the emissions.

If one used a simple intuitive approach to the example, it would appear that the facility does not need to obtain any ERCs, because its emissions were permitted for 585 tons per year in 1985, and the emissions reductions since then have been a total of 104 tons per year, more than sufficient to cover the 70 tons per year needed for the new process.

However, there are four problems with that intuitive rationale:

1. The 15-ton-per-year emission reduction apparently obtained by shutting down Large Degreaser No. 2 in 1992 is not permanent, because the permit was not surrendered. Therefore, that emission reduction is not available for offsetting the new process.
2. The 10-ton-per-year emission reduction that occurred in 1986 also does not count, because it occurred more than 5 years prior to the new construction in 1994.
3. Only 6 of the 60 tons per year related to the shutdown of XYZ Process No. 1 can count for offsetting, because 90 percent of the original 60 tons per year were required reductions per the local permitting authority's SIP.
4. The emissions reduction of 4 tons per year associated with the unpermitted degreaser does not count, as it was not permitted.

Therefore, the facility must acquire 49 tons per year of ERCs to accommodate the 70 tons per year of new process. This value is calculated as follows:

> New process increase of 70 tons per year minus the 15-ton-per-year emission reduction obtained by shutting down Large Degreaser No. 1 minus 6-ton-per-year surplus emissions from XYZ Process No. 1 equals 70 minus 21, or 49 tons per year.

One interesting aspect of the above example is that if the emissions reduction from the shutdown of Large Degreaser No. 1 was equal to 25 tons per year or more, the net emissions increase would be 70 minus 6 minus 25 tons per year, or 39 tons per year. In this case, the modification would not be "major," and at least from a federal perspective, no ERCs or other nonattainment requirements would apply.

Value of ERCs

ERCs are valuable commodities, because new sources cannot be permitted in nonattainment areas without obtaining ERCs. It is generally considered that the value of ERCs is the marginal cost of controls.

For example, assume that it costs a facility $1,000,000 to install a retrofit control technology that would reduce emissions by 500 pounds per day. Further assume that the facility wishes to expand operations and increase emissions from another device by 500 pounds per day. In this case, the facility is faced with either purchasing ERCs or installing the retrofit controls as part of an internal offset (netting) strategy. If the ERCs cost less than \$2000 per pound per day to purchase from a third party (i.e., \$1,000,000 divided by 500 pounds per day equals \$2000 per pound per day), it will be less costly to purchase ERCs. On the other hand, if ERCs cost \$4000 per pound per day to purchase, then the facility owner will opt to install the retrofit controls, because the cost of the ERCs will be twice as great as the cost of retrofit controls.

When one is evaluating the cost of ERCs, the marginal cost of retrofit controls in terms of cost per emissions unit (i.e., dollars per pound per day) is a useful starting point. However, just as in any market, the law of supply and demand can significantly affect the values of the ERCs to any specific source. In addition, one must consider the fact that, regardless of the availability of ERCs, LAER must still be installed. Therefore, the amount of ERCs needed is often less than 5 percent (i.e., at least 95 percent control) of the uncontrolled emissions.

ERC Banking

One of the techniques used to avoid losing ERC credits is to apply to the local permitting authority for an ERC certificate. The ERC certificate, once obtained, is valid indefinitely until used to offset new emissions. This is termed *ERC banking*. The application procedure for an ERC certificate is relatively straightforward, but the various regulatory concepts that reduce the magnitude of the ERCs (i.e., real, quantifiable, permanent, surplus, discounting) need to be taken into account. Therefore, depending on the area, there may not be any ERC certificates on file. In addition, since a facility often does not know whether or not it will need the reductions for internal offsets, ERC trades are not as common as one would think, thereby further reducing the number of ERC banking applications submitted and issued.

NET AIR QUALITY BENEFIT

One aspect of permitting new sources in nonattainment areas is the requirement that there be a net air quality benefit when the new source is permitted. Some facility operators assume that if offsets are obtained at a ratio greater than 1 to 1, then there will be an air quality benefit. That would be the case if the emissions reductions affected the same area and with the same pollutants as the emissions increases. However, this equivalence occurs only when the reducing and increasing sources are located at exactly the same place and have exactly the same emitting characteristics.

Therefore, in order to meet the "net air quality benefit" test, one may need to perform dispersion modeling to show that the reductions more than offset possible ambient air quality impacts of the proposed source. In practice, for ozone, these demonstrations can often be done qualitatively rather than through a detailed digital air shed model. For the nonreactive pollutants, such as CO and SO_2, these demonstrations would normally require traditional Gaussian dispersion modeling (Chapter 4).

In addition to the net air quality benefit analysis, the 1990 CAA also requires an alternative sites analysis to demonstrate that the benefits of locating in the nonattainment area outweigh the environmental and social costs of the project. This alternatives analysis must include not only looking at physically different sites for the new project, but also looking at alternative processes that emit less pollution, or different sizes of plants (e.g., building a smaller plant in the nonattainment area as a distribution point and building a larger plant outside the nonattainment area).

REQUIRED POLLUTANT-SPECIFIC NONATTAINMENT PROGRAMS

Depending on the severity and the nature of the nonattainment pollutant, the 1990 CAA mandates certain minimal programs that must be contained in the SIP. These programs are as follows. Note that there are three additional issues that relate to the nonattainment classification status of an area, other than the measured design value.

One item is that the EPA can "bump down" (i.e., make less severe) or "bump up" (i.e., make more severe) an area classification if the design value is within 5 percent of the lower or upper classification. (Interestingly, from a practical standpoint, monitored data are, at best, only plus or minus 15 percent accurate.) The second issue is that a state can bump up an area's classification voluntarily. The advantage of a bump up is more time for compliance; the disadvantage is the severity of the mandatory minimum requirements. The third is that the EPA can give two one-year extensions to the attainment deadline.

Ozone

Marginal Areas

1. Motor vehicle inspection programs must be adjusted to become more rigorous. In addition, current EPA policy prohibits inspection and repair from occurring at the same facility, contrary to many state or local programs.
2. RACT must be implemented in accordance with the Control Technique Guidelines (CTGs) published by the EPA. Previously, what was "reasonable" under RACT was more open to interpretation. This means that a facility should track the progress of CTG development for its source category, since the conclusions in that CTG could have far-reaching consequences.
3. Offset ratio is 1.1 to 1.
4. All sources of VOC and NO_x emissions greater than 25 tons per year must submit annual emission statements (even if the source is not a Title V major source; see Chapter 12).

Moderate Areas

1. All requirements for marginal areas.
2. Stage II vapor recovery for gasoline dispensing stations (i.e., the vapor recovery nozzles at the gasoline pump common to California and many other major metropolitan areas) must be installed for stations dispensing more than 10,000 gallons of gasoline per month.
3. SIP must include a detailed plan of annual emission reductions to meet the six-year attainment deadline.

Serious Areas

1. All of the marginal and moderate area requirements.
2. Enhanced ambient air quality monitoring requirements to better determine the area's air quality.
3. Major source threshold reduced to 50 tons per year, offset ratio increased to 1.2 to 1.
4. Special provisions for modifications to existing sources:
   a. *De minimis* level changed to 25 tons per year.
   b. Sources less than 100 tons per year need to install BACT.
   c. Modification trigger can be avoided if a 1.3 to 1 offset is used and source emits less than 100 tons per year.
   d. Sources over 100 tons per year can avoid LAER if the offset ratio is increased to 1.3 to 1.

5. Enhanced vehicle inspection and monitoring program. (The enhancements include adding computers to the inspection and monitoring test equipment, providing for denial of vehicle registrations if a vehicle owner does not repair the vehicle, separating testing and repair facilities, and providing for higher repair waivers.)
6. Required clean fuels program or alternative to meet the same goals as a clean fuel program. (Clean fuels include natural gas, methanol, or ethanol.)
7. Three percent per year annual reduction or demonstration that all feasible measures are being employed.
8. Demonstration on a three-year basis that actual vehicular data are consistent with projections in SIP.

Severe Areas

1. All requirements of marginal, moderate, and serious areas.
2. Major source threshold 25 tons per year.
3. Offset ratio 1.3 to 1 or if *all existing* major sources install BACT, then only 1.2 to 1.
4. Transportation control measures to offset increased vehicle miles traveled, including mandatory employer trip reduction requirements.

Extreme Areas

1. All requirements of marginal, moderate, serious, and severe.
2. Major source threshold 10 tons per year.
3. Offset ratio 1.5 to 1 unless all existing major sources install BACT, then 1.2 to 1.
4. Any amount of VOC increase is treated as a major modification, but external offsets are not required if internal offsets exceed a 1.3 to 1 ratio.
5. By 1992, commercial boilers must burn clean fuel 90 percent of the time unless they emit less than 25 tons per year.
6. SIP may establish traffic control measures.

Interstate Transport Commissions Recognizing that ozone formation is a long-range transport phenomenon (i.e., emitted pollutants require time in the atmosphere to react and form ozone), the 1990 CAA established a Northeast Ozone Transport Region and provided for establishment of additional interstate transport regions as necessary. These regions are established when interstate transport of pollution is a significant contributor to nonattainment in any one state.

The Northeast Ozone Transport Region contains 11 states from Maine through Maryland and parts of Virginia and the District of Columbia. In these regions, Stage II vapor controls at gasoline dispensing stations are required and RACT must be installed for sources over 50 tons per year of VOCs.

One other aspect of these regions is that, if there is a SIP call for a particular state, all of the states in the transport region must also resubmit their SIPs to address the regional nonattainment issues.

Carbon Monoxide

The following provisions for carbon monoxide are the same as for ozone:

1. The "within 5 percent of design value" reclassification
2. Two one-year extensions to achieve attainment

Moderate Areas

1. Inventory of all CO sources must be submitted and updated every three years.
2. Annual forecast of vehicle miles traveled (VMT) through the attainment date must be submitted to the EPA.
3. A contingency plan must be implemented if actual VMT exceeds annual forecasts.
4. The vehicle maintenance and inspection program must be upgraded.

Serious Areas

1. All provisions of moderate areas.
2. Transportation controls measures (TCMs) must be proposed and implemented.
3. Oxygenated fuels program must be implemented if needed to achieve attainment.
4. Major source threshold reduced to 50 tons per year *if* major sources significantly contribute to nonattainment.
5. If needed, states must submit a plan to include economic incentives (e.g., marketable permits, auctions of emission rights) and transportation planning in order to ensure attainment by the deadline.

Particulate Matter

1. Two one-year extensions of attainment date can be granted.
2. May also obtain a five-year extension if it is impractical to achieve attainment and the SIP contains measures at least as stringent as all other states.

(This provision is to account for unusual meteorological conditions that make attainment very difficult.)

3. If man-made (anthropogenic) sources of emissions do not contribute significantly to nonattainment, the EPA can waive the SIP requirement and/or redesignate the area attainment despite measured exceedances of the standard.
4. For serious areas of PM_{10} nonattainment, the major source threshold is reduced to 70 tons per year, and the SIP was required to incorporate best available control measures (BACM) by November 15, 1994. BACM must include measures to address urban fugitive dust (e.g., from construction activities), residential wood combustion, agricultural burning, and silviculture activities.
5. Five percent per year annual emissions reduction demonstration is required. The emissions include PM_{10} and PM_{10} precursors (e.g., sulfur oxides that react with other pollutants, such as ammonia, to create PM_{10}, ammonium sulfate).

SANCTIONS FOR NOT SUBMITTING AN APPROVABLE SIP OR NOT REACHING ATTAINMENT

The 1990 CAA contains a number of sanctions that will be applied to the state if the state does not submit an approvable SIP in accordance with the various due dates and/or the nonattainment areas do not come into attainment by the required dates. (An "approvable" SIP is one that contains at least the required elements and will cause the area to come into attainment by the deadline.)

The sanctions take a variety of forms, but include one or more of the following:

1. Bump up in the severity classification, regardless of the design value.
2. Severe and extreme ozone nonattainment areas must collect penalty fees on major sources equal to $5000 per ton of VOC emissions per year above 80 percent of the "baseline amount." The baseline amount is the annual actual or allowable emissions, whichever is lower.
3. Ban on new highway or sewer construction funds unless the funds go to safety or pollution reduction-related projects.
4. Increase emissions offset ratio to 2 to 1 for the affected nonattainment area.

INNOVATIVE CONCEPTS USED TO ACHIEVE ATTAINMENT

Because of the difficulty of achieving attainment, many states and local permitting authorities are attempting various innovative concepts to reach attainment. Some of these concepts are even specified by the 1990 CAA. The innovative concepts include, but are not limited to,

1. Marketable permits
2. Auctions for emissions rights
3. Open market trading
4. Market incentive reductions
5. Interpollutant trades
6. Mobile/stationary source emission trades
7. Community banks of ERCs

Market Incentives

The first four programs are generally referred to as market or economic incentives and are best exemplified by the South Coast Air Quality Management District (SCAQMD) Regional Clean Air Incentives Market (RECLAIM) that was recently (October 1993) enacted in the Los Angeles Basin (SCAQMD, 1993). The RECLAIM program has as a fundamental tenet that, if industry is given the freedom to choose the best way to achieve emission reductions, it will naturally choose the most cost-effective method. In this way, it is thought that industry will achieve the reductions more efficiently and completely than would occur under the old "command and control" system of air quality regulation.

The RECLAIM program currently applies to all sources of NO_x and SO_2 in the SCAQMD that emit more than 4 tons per year of either pollutant. (VOCs may be included in RECLAIM at a later date, but the program began with only SO_2 and NO_x.) At the beginning of RECLAIM, each facility is given a baseline allocation of emissions. The baseline was determined through a historical review of the previous five years of emissions and choosing the highest annual emission rate based on throughput of fuel or process weight. Then, all currently applicable rules that were to reduce emissions were applied to the baseline. This gives the facility starting point. The District then summed all of the starting points to determine the total emissions in the RECLAIM universe.

Using the various predictive models and AQMP requirements, the District then estimated the total amount of emissions that could occur when the

District is attainment at the deadline date (2001 for NO_x). This amount of emissions is the end point. The difference between the total starting point and the total end point, expressed in terms of percent reduction per year, is then mandated on all RECLAIM sources. In other words, each source must reduce its emissions by the same percentage each year.

When the starting and end points for each source are established, the District steps back and simply monitors how the source complies with the required reductions. The monitoring and reporting requirements are quite complex and essentially require every source to install continuous emission monitors (CEMs) that automatically electronically report emissions data on a real-time basis to the District.

To meet the required reductions, a facility is allowed to use one or more of the following techniques:

1. Apply a facility bubble and use internal offsets (offset ratio 1 to 1).
2. Shut down equipment or reduce throughput capacity of equipment.
3. Apply emission controls or replace equipment with less polluting, newer equipment.
4. Purchase RECLAIM Trading Credits (RTCs). RTCs are just like ERCs when a separate facility reduces emissions by an amount greater than required, thus making the additional reductions available for transfer to other facilities. The major difference between RTCs and ERCs is that the RTCs are temporarily issued on a quarterly basis. Therefore, a facility that is experiencing a slowdown in production in one quarter can temporarily transfer their emission allowances (RTCs) to another facility that is increasing production.

Open Market Trading

The RECLAIM program and most market incentive programs to date have been of the "emissions budget" or "cap and trade" variety. Under these systems an emissions baseline is established, emissions are then capped at some level equal to or less than the baseline, and then emission trades are allowed to occur beneath the cap. The 1990 CAA Title IV acid rain provisions and RECLAIM are examples of cap and trade or emissions budget market incentive programs. In these programs, emissions are traded permanently or semipermanently.

Another type of market incentive program is the "open market trading" type. The EPA has recently (August 3, 1995; 60FR3961) promulgated a proposed model Open Market Trading Rule (OMTR). OMTR was first used in the

Northeast U.S. Ozone Transport Region. The OMTR program does not require establishment of an emissions baseline (a time-consuming and very contentious process) or an emissions cap. Rather, the OMTR allows a facility to reduce emissions by whatever means feasible to create a discrete emission reduction, or "DER." DERs are expressed in terms of tons of emissions. DERs are then sold to other facilities to use as they need to meet other emission limits. Once the DERs are used, they are gone.

The distinction between "emissions budget" and "OMTR" programs is subtle but important. In OMTR, emissions are traded on a one-time-use basis. The unit of trade is tons of emissions. In emissions budget programs, emissions are traded at least semipermanently (e.g., on a quarterly or annual basis) and the unit of trade is expressed in an emission *rate*—for example, tons of emissions *per year*. OMTR programs are more flexible, are easier to implement, and take full advantage of natural market forces.

Interpollutant Trades

As previously noted (Chapter 3), some pollutants react in the atmosphere to form other pollutants. For example, NO_x can react to form ozone, SO_2 can react to form particulates, and VOCs can react to form ozone and/or particulate. Therefore, in some jurisdictions, interpollutant trades are allowed as part of the overall attainment strategy. For example, reductions in SO_2 can be used to offset increases in particulates or NO_x reductions can be used to offset increases in VOC.

Because the science of chemical transformations in the atmosphere is still under study, the precise quantities traded are often subject to considerable controversy, and rarely is a one-for-one by-weight trade allowed.

Mobile/Stationary Source Trades

Another concept is to use mobile source reductions to offset stationary source increases. Mobile source reductions can occur from either scrapping old cars that tend to pollute more or implementing a ride-sharing or other program to reduce the amount of vehicular traffic. One of the major difficulties associated with mobile to stationary source trades is the difficulty of demonstrating that the mobile source reductions are real and permanent.

Community Banks for ERCs

In those jurisdictions where there is a zero threshold for major sources (i.e., any amount of emissions must be permitted and subject to the regulatory and offset program), there has been established a community bank of ERCs. ERCs from the community bank are available at essentially no charge and are used to offset emissions from new sources or modifications that increase emissions at existing sources. The community bank is "funded" from the overall emission reductions built into the SIP. The reason for a jurisdiction establishing a zero threshold and then "giving back" ERCs to sources without charge is to force *all* sources in the jurisdiction (large or small) to install LAER prior to construction.

Summary

There are a number of additional control systems and other concepts embodied into SIPs and the facility environmental manager needs to carefully track those concepts as they affect facility operations. For example, in the RECLAIM situation, the District's baseline emission inventory was often incorrect for facilities. Those facilities that did not participate in the development of the program were inadvertently adversely affected.

CHAPTER

10

PERMITS AND THE NEW NATIONAL OPERATING PERMITS PROGRAM

The 1990 CAA established a new major permitting program, termed the *national operating permit.* This program is estimated by the EPA to affect approximately 34,000 major sources and an additional 350,000 nonmajor sources nationwide. The program is complex and significantly expands air quality permitting.

PERMITS PRIOR TO THE 1990 CLEAN AIR ACT

Prior to the 1990 CAA Amendments, generally two types of permits were required by the local permitting authority (i.e., the state, local air quality control agency, Native American tribe, or EPA). The two types were the permit to construct (PTC) and the permit to operate (PTO). In many jurisdictions, only the PTC was required and, once the PTC was obtained, the source was constructed and operated generally in accordance with the PTC and local regulations.

In the more sophisticated jurisdictions, a PTO was also issued after the source was constructed, but before it could routinely operate. In these jurisdictions, the PTO was essentially automatically issued after a cursory inspection of the constructed source. The purpose of the inspection was to verify that the source was constructed and capable of operating in accordance with the PTC. (Verification of operations usually required a source emissions test to verify emission rates.) In even the most advanced jurisdictions, however, the PTO was usually only a few pages long with no more than about 15 different conditions associated with the permit. After a PTO was obtained, a copy of it was required to be posted on or near the equipment for which the permit was issued.

Another aspect of these PTCs and PTOs that bears noting is that the permits were generally issued for individual devices at a facility, not for the entire facility as a whole. These types of permits have been characterized as being similar to an elevator permit: Obtain it and post it.

Permit Conditions

Even in this relatively simple permitting environment, how permit conditions were constructed and written was important. A carefully crafted permit application could yield significant dividends when the permitting authority used the application to specify final permit conditions. A poorly crafted application, likewise, often resulted in permit conditions that were impossible to follow and caused noncompliance issues even though the equipment was being operated in accordance with the intent of the application.

In previous chapters, various types of permit conditions have been discussed. These conditions and typical errors will be summarized herein. The purpose of examining typical permit conditions under the existing "elevator permit" type of permitting is that the new Title V permitting program will extend the concept of permit conditions to a new and severe level of complexity.

Emission Limits One of the most common errors is to specify emission limits derived from assumed maximum control efficiencies of control equipment. For example, a device may be 95 percent efficient when the inlet loading of a pollutant is 100 pounds per hour, but 98 percent efficient when the loading is 300 pounds per hour.

If the permit condition reads that emissions from the device shall be no more than 6 pounds per hour, it may appear to be acceptable. (That is, 98 percent control means that 2 percent of the inlet emissions are released from the device, and 2 percent of 300 is 6 pounds per hour. Likewise, when control is 95 percent, 5 percent of the emissions are released, and 5 percent of 100 is 5 pounds per hour.) However, a careful reading of the vendor's performance guarantee leads one to ask what the control efficiency is when the inlet loading is between 100 and 300 pounds per hour. If the equipment operates at only 95 percent control until the inlet loading reaches 200 pounds per hour, then ramps up to the 98 percent level, there could be a compliance problem, in that 95 percent control at an inlet of 200 pounds means that 10 pounds per hour is emitted, a value greater than the permit condition.

There are similar problems if the permit condition reads that the emissions from the device shall be no greater than 6 pounds per hour and 98 percent removal maintained. This problem is obvious, because the equipment vendor only guaranteed the 98 percent removal efficiency at an inlet of 300 pounds or greater.

There are essentially two ways to get around this problem. One is to overdesign the equipment, and the other is to be very precise in writing the permit conditions to allow for flexibility.

Time Period Averaging for Permit Conditions In this author's experience, rarely are time periods for averaging emission tests or performance results provided in permits. The averaging time then becomes an interpretation issue or defaults to a time period specified in a test method. Whenever a permit condition is open to interpretation, the facility manager has a potential compliance problem. Therefore, specifying a time period for emissions limits is important. In the above illustration, for example, the permit condition should read "Emissions no greater than 6 pounds per hour averaged over a three-hour interval and 98 percent removal efficiency when the inlet loading is 200 pounds per hour or greater."

Vendor and Engineering Guarantees Equipment vendors usually have an engineering and operational focus, not a compliance focus. Therefore, when purchasing a new piece of equipment, it is critical to include in the contract for the equipment terms that force the vendor to meet not only its normal operational and engineering performance specifications, but also the compliance requirements of the local permitting authority. One should analyze the regulatory requirements for the piece of equipment, develop draft permit conditions, obtain informal or formal regulatory review of those conditions, and then include them in the contract for the piece of equipment. In addition, do not release the contractor from the job until he has demonstrated performance with respect to the regulatory terms as well as the traditional engineering and operational performance terms. Many facilities will use a pre-compliance test conducted internally with the vendor present prior to requesting compliance certification from the local permitting authority. At this precompliance testing phase, the vendor can adjust the equipment and fine-tune it with respect to the compliance requirements. In this way, potential noncompliance penalties and difficulties with the local permitting authority can be avoided.

Source Emissions Testing When source emissions testing (often referred to as *stack testing*) occurs, if it is for demonstration of compliance to a regulatory agency, the testing should be done in accordance with published and preapproved source testing protocols. Although there are a number of published testing protocols, there is not a national standard protocol for all pollutants or source types. In addition, the local jurisdictions often have their own protocols that are modifications of EPA protocols and/or include pollutants or source types that the EPA does not.

One must be careful when performing testing in accordance with a local jurisdiction's protocol. If the source is regulated only by state regulations (i.e., the local regulations are not part of the SIP and are not required by the EPA), then performing a source test only in accordance with local protocols will suffice. If, however, the source is also federally regulated (this type of source is said to have "federally enforceable" limitations), then the testing must also meet EPA protocols, or the test may not be deemed valid by the EPA. This type of problem most often occurs when one is performing a source test for NSPS compliance demonstration.

Most local jurisdictions simply incorporate federal NSPS without further elaboration. Because NSPS is a federally mandated program, strict compliance with the NSPS is required. NSPS standards often include detailed source emissions testing requirements to demonstrate compliance. If the source testing is performed in accordance with local protocols, and the EPA does not consider those protocols to be equivalent to EPA protocols, the source test will not be considered valid, and the EPA can cite the source for noncompliance, even though it complied with the local jurisdiction's testing requirements.

This issue highlights a common problem that has existed with air quality permitting in the past. A source can be complying with all of the conditions specified in its permit, but not necessarily be complying with all of the other regulatory requirements of the EPA and the local jurisdiction. This issue has been discussed in Chapter 3, but it is important to bring up here again. (In the new national operating permit program, this issue will be resolved through the permit shield.)

Summary

The above describes just a few simple examples of permit condition problems. Almost every permit has interpretation difficulties. Therefore, it is very useful in the permit application to suggest permit conditions. If suggested conditions are not included, at least state the emissions and operational data in such a manner that the permitting authority can easily write permit conditions from the application. One must remember that the permitting agency does not have time to thoroughly study the permit application. Many jurisdictions process thousands of permits per year with limited staff. Therefore, there must be a great deal of up-front planning on the part of the applicant in order to help ensure that the permit is "livable."

PERMITS AFTER THE 1990 CAA

The 1990 CAA has significantly changed the nature of permits for facilities. Although some jurisdictions may still require a PTC prior to construction (the EPA's PSD and NA programs for major sources do require such permits; see Chapter 9), the PTO has now become a significant permit document that specifies not only the maximum allowed emission rates, but also exactly how the facility will be operated and how certification of compliance will be determined. The new permits are variously referred to as national operating permits, Title V permits (since the legislation creating the program is contained in Title V of the 1990 CAA), Part 70 permits (since the federal permit regulations are contained in 40 CFR Part 70), or major source permits (since the new program initially applies only to major sources). In this document, the term "Title V" permits will be used.

Major and Minor Source Permitting

As has been previously explained, both major and minor sources usually require a permit, although there is usually also a *de minimis* level of emissions below which no permit for devices is required. The initial Title V permitting program applies only to major sources (as defined by Title V); however, local jurisdictions can add nonmajor sources to the program as they desire. Major sources are defined specifically for the Title V program and include traditional major sources (as defined under NSR; see Chapter 8) as well as some sources previously considered to be minor sources (e.g., HAPs sources, Chapter 7).

Title V major sources include those that

1. Emit more than 100 tons per year of any regulated pollutant (both criteria and non criteria) in attainment or nonattainment areas
2. Emit more than the nonattainment major source thresholds of nonattainment pollutants in nonattainment areas (as noted in Chapter 9, these thresholds vary from 10 tons per year for VOCs in extreme ozone nonattainment areas to 100 tons per year for other pollutants in other nonattainment areas)
3. Emit more than 10 tons per year of any single federal HAP (Chapter 7) or 25 tons per year of aggregate federal HAPs
4. Are subject to New Source Performance Standards (NSPS, Chapter 6)
5. Are subject to National Emission Standards for Hazardous Air Pollutants (NESHAPs, Chapter 7)
6. Are any other source category that the EPA or a state explicitly includes in the permitting program

The EPA has estimated that 34,000 major sources will be affected by the Title V program. Of these 34,000 sources, it is estimated that 62 percent are triggered by the 100-ton-per-year threshold, 27 percent by nonattainment area thresholds, and 11 percent by air toxics becoming regulated at the 100-ton-per-year level (ABA, 1993). The Agency also estimated that as many as an additional 350,000 minor sources could come into the program as a result of the air toxics MACT program (i.e., the NESHAP trigger for the 10- or 25-ton-per-year threshold).

Fugitive Emissions To the extent that fugitive emissions are quantifiable, fugitive emissions count in Title V determinations for all sources of HAPs. However, for non-HAPs emissions, fugitive emissions count only for the 28 major emitting facilities (MEFs, Chapter 8).

Regulated Pollutant The 100-ton-per-year threshold for Title V major sources is based on "any regulated pollutant." This term means any pollutant regulated by the 1990 CAA, which is a much broader term than the traditional definition of major source which is for "criteria pollutants" only.

The EPA has published guidance on the term "any regulated pollutant" (USEPA, 1993). Simply stated, the term includes the following categories of pollutants:

1. Nitrogen oxides and volatile organic compounds.
2. All criteria pollutants: sulfur dioxide, ozone, nitrogen dioxide, carbon monoxide, PM_{10}, and lead.
3. Any pollutant subject to New Source Performance Standards (NSPS) promulgated by the EPA, which includes pollutants for which emissions standards, work practices and standards, or other requirements have been issued. This list of pollutants is the same as in the PSD modification threshold determination shown in Chapter 8 and includes such substances as total reduced sulfur compounds, fluorides, total suspended particulates, as well as PM_{10} and some metals.
4. Any of the ozone depleting chemicals (ODCs) specified as either a Class I (CFCs, methyl chloroform—1,1,1-TCA) or Class II (HCFCs) substance (Chapter 12).
5. Any pollutant subject to a standard promulgated under the Title III HAPs program of the 1990 CAA. This includes the list of 189 HAPs (for which NESHAPs have been promulgated) and the list of chemicals for which accidental release prevention is required (Appendices I and J).

There is a slight technicality related to when HAPs are considered "regulated pollutants." The pollutant does not become regulated until a NESHAP is promulgated. Since the HON rule has been promulgated (Chapter 7), 149 of the 189 HAPs are included in that rule, and therefore there are only another 40 HAPs for which a standard has not been promulgated as of this date. Technically speaking, if a NESHAP for a pollutant has not been promulgated, the HAP is not a regulated pollutant. However, because many more NESHAPs for the 189 HAPs will be promulgated, from a practical perspective, one should assume that all 189 HAPs are regulated pollutants.

One other item important to note is that when a source becomes a Title V major source and requires a Title V permit, *all* devices and regulated pollutants emitted by the source must be included in the permit, even if the emissions are less than the major source thresholds. There is one exception to this: if the local permitting authority exempts the listing as either an "insignificant activity" or a "*de minimis* emission." These determinations are local permitting authority issues, and the EPA only provides that such exemptions should be allowed in the local permit regulations.

TITLE V PERMITS

The national operating permit program is being developed during the period from 1990 through 1996 and beyond. The concept for the operating permits is similar to the National Pollution Discharge Elimination System (NPDES) permits issued to facilities under the Clean Water Act. The fundamental concepts of these types of permits include the following:

1. There is a five-year term for the permit.
2. The permit applies to the entire facility ("pollutant emitting devices" for air quality and "outfalls" for water quality);
3. Routine compliance certification to the agency is required.
4. Once a permit is issued, if operations continue in accordance with the permit, there will not be enforcement citations.
5. Public participation in review of the permit is encouraged.

Title V Regulations Implementation Schedule

EPA Title V regulations were published in the Federal Register on July 21, 1992, and are codified in 40 CFR Part 70. The states had until November 15, 1993, plus a one-year extension, to formally propose regulations implementing Part 70 at the state and local level. The Part 70 EPA regulations allowed the states considerable flexibility. Therefore, there will be and are major differences among various state programs. If a state has local jurisdictions with air quality delegated authority, the local Title V regulations must also have been promulgated by the state deadline.

The state regulations were to be submitted to the EPA by November 15, 1993, and the EPA then has one year to review and either approve, partially approve, or disapprove the state's program. Unless the state's program is deficient in obvious and major ways, it is not expected that a state program will be completely disapproved, and the EPA has been issuing interim approval to most local jurisdictions. Rarely has complete approval been given, and the regulations are constantly changing.

When the state's program is approved, Title V sources supposedly have 12 months in which to submit their Title V permit applications. Note that the trigger is one year from the approval date. However, there is nothing prohibiting a state from requiring a Title V permit application early even if the state's program has not yet been fully approved by the EPA. Several states are, in fact, planning to require permit applications well before the deadline date.

TABLE 10-1 General Time Frames for Title V Operating Permits

Associated Action or Program Element	Key Date or Time Frame
Overall time frame for permitting agency to issue or reissue permits and significant modifications	18 months after submittal of the application
Overall time frame for permitting agency to issue a minor permit modification	90 days after submittal of the application
Overall time frame for an administrative amendment	60 days after submittal of the application
Overall time frame for a reopening by the permitting authority to incorporate new or revised applicable requirements	30 days' advance notice to applicants; reopening is completed 18 months after the regulations take effect

The states have one year after their permit program has been approved to review and approve or disapprove the first-third of the applications. The states then have another year to act on the second-third and a third year to act on the last third of the applications. Therefore, all permits must be issued no later than three years after the state program has been approved—that is, no later than November 15, 1998. Because the permits will be complex and difficult to issue and approve, many states are requesting permit applications prior to EPA approval of the program in order to get a head start on the review process. Practically speaking, this means that a source will be preparing a permit application in accordance with regulations that will almost certainly change during the application preparation and review process.

Title V Permit Application Review Process

The review process under Title V is detailed and involves several steps. Table 10-1 shows the general time frames for permit application review, issuance, and renewal. It also indicates a time line for permit revisions (to be discussed in the next sections).

Table 10-2 shows the specific time frames for the permit application review process. It is obvious from this table that the one year allotted to issue the first one-third of the permits is a very short time frame given the amount of public review and other agency actions required.

The following sections discuss the various aspects of the permit application and review process. The topics covered include

TABLE 10-2 Time Periods Within the Review Process for Title V Permits

Associated Action or Program Element	Key Date or Time Frame
Notification to permit applicant on the completeness of the application	60 days following receipt of the application
Public hearing on all permit issuances, renewals, and significant modifications	Must allow for 30 days' notice to public and affected states
EPA review for all permit issuances, renewals, and significant modifications	45 days after receipt of draft permit
Time period for state to correct deficiencies following an EPA disapproval	90 days after EPA disapproval of permit
Time period for public petition to EPA relevant to state or EPA permit decision	60 days after end of EPA 45-day comment period
EPA approval or denial of the public petition	90 days following submittal of the petition to EPA
Judicial review of permit issuance decision in state courts	90 days after permit issuance unless new grounds for challenge arise

1. Required elements of the permit application
2. Operational flexibility
3. Permit modifications
4. Permit reopeners
5. Public notice and review
6. Permit and application shields
7. Annual emission fees
8. Annual compliance certifications

Required Title V Permit Application Elements

The following elements are required to be included in the Title V permit application:

1. *Company Information.* Facility name, location, owner, plant manager, and other officials responsible for the permit, including the name of the responsible corporate official who will be signing the annual certification of compliance.

One of the problems some companies have is that the environmental manager is in a central location and he or she manages several different facilities. Therefore, the environmental manager puts the same address on all applications, even though the different facilities are identified by name. This confuses the local agencies and often leads to apparent recordkeeping violations (because information filed with the agency is filed with the wrong facility). To avoid this problem, one should clearly identify the different facility addresses and list a responsible official *at the facility* as the addressee. It is easier to use the company's internal mail system to distribute the permitting and enforcement information than to rely on the agency to keep different facility names straight.

2. *Plant and Emitting Devices Description.* The application must include a description of the processes and emitting devices at the plant. Along with the description, there must be an analysis of the emission quantities and types, as well as a calculation of the maximum potential to emit. Operational scenarios that yield different emissions must be detailed, and a demonstration of how the emissions are calculated must be provided.

The Title V permit is a facilitywide permit. Therefore, the total emissions of the facility must be detailed, not just the individual devices. On the other hand, it is not generally acceptable to list only the total emissions without indicating what emissions will occur from the individual devices. The term "facility" is the same as under New Source Review (Chapter 8) and includes all devices located on contiguous or adjacent properties that are under common control and ownership and are related to the same industry type. The facility bubble is the key to operational flexibility.

When detailing emission rates of the facility under various scenarios, it is important to note that emission rates for individual devices that have been previously established by the former device-type of permits or by local rules or regulations cannot be violated when developing the operational scenarios. If the limitations are *only* related to local requirements that are not part of the SIP, then the limitations *may* be able to be violated under operational scenarios, depending on local agency discretion.

3. *Emissions-Related Data.* The facility must provide product throughput, feed rates, raw materials type, usage rates, and other data necessary to demonstrate the amount of emissions.

4. *Air Pollution Control Equipment.* Any control systems that are in place must be described, and evidence indicating the control efficiency must

be provided. In addition, one must be careful to identify emissions that the control systems themselves may create. For example, afterburners (sometimes referred to as thermal oxidizers) that are used to control VOC emissions through combustion of the VOCs generate NO_x emissions through the combustion process. In addition, natural gas combustion generates benzene and formaldehyde emissions, which are federal HAPs and, therefore, must be listed unless exempt through the *de minimis* exemptions.

5. *Monitoring, Recordkeeping, and Reporting Procedures.* These procedures must be detailed, including the frequency of reporting, the method of reporting, and monitoring. Specific testing methods must be detailed, as well as the goal of the test.

6. *List of Insignificant Activities.* This list will identify those emitting devices and processes that do not have significant emissions, as defined by regulations promulgated by the local permitting authority.

7. *Description of Applicable Requirements and Nonapplicable Requirements.* The facility has a permit shield only for those requirements of the Act that are applicable and those that are not applicable as listed in the permit application. This list of requirements should include not only the requirement (e.g., "Regulation 400, Rule 401"), but also what the requirement is and how it applies or does not apply to the device in question (e.g., "Regulation 400, Rule 401 limits emissions of VOCs to no more than 10 pounds per day from paint booths of less than 10,000 scfm flow rate. Paint Booth XYZ meets the requirements of Regulation 400, Rule 401, since its maximum potential to emit is 8 pounds per day.")

8. *Treatment of Confidential Information.* No part of the permit can be kept confidential. Therefore, if there is trade secret or other confidential information that is relevant to emissions or other permit conditions, the information must be kept out of the permit and alternative arrangements made with the local permitting authority.

9. *Compliance Plan.* The permit must contain methods to demonstrate "continuous compliance." This does not necessarily mean that the facility must install and operate continuous emissions monitoring equipment (CEMs), but that there must be a system of evaluating compliance. "Surrogate" emissions monitoring parameters can be specified and should be developed as part of the permit application. One example of surrogate parameters is to monitor the normal operating conditions for a boiler, such as combustion temperature, percentage of excess air, and fuel flow. If past source emissions testing can be shown to correlate well with those parameters, then the local

permitting authority should be able to accept the surrogate parameters instead of CEMs to demonstrate compliance.

Operational Flexibility

The Title V permit is an *operating* permit. Therefore, it describes how the facility will be operated over the next five years. Deviations from the description in the permit are not allowed except for certain circumstances. There are three general types of operational flexibility: (1) those allowed by the permit, (2) those allowed only after permit modification, and (3) those allowed off-permit (i.e., they are not addressed in the permit).

The kinds of operational changes that can be made without requiring a permit modification are as follows:

1. The changes were anticipated and listed as alternative operational scenarios in the permit itself. Implementation of any one of these scenarios without notice to the agency is allowed at any time.
2. Emissions trading within the facility occurs, even though not listed as an operational scenario. This is allowed with a seven-day advance notice to the local permitting authority, as long as the trading does not violate any of the applicable underlying requirements.
3. Emissions trading within the facility occurs, and it is listed in the permit. This is essentially the same as an alternative operating scenario and is allowed without notice.
4. Emissions trading occurs within an "independent" emissions cap. An independent emissions cap is a facilitywide limit on emissions that is not required by an applicable underlying requirement (contrary to item 2 above) but has been voluntarily accepted and listed in the permit. These changes can occur without notice.
5. The changes are "off-permit" in that they are not proscribed or prohibited by the permit. This type of change would be allowed as long as it does not violate an applicable underlying requirement. Off-permit changes require advance notice to the local agency.
6. Changes that contravene general provisions of the permit. These changes require seven-day advance notice and may not contravene any applicable requirement or federally enforceable permit term. The most commonly cited example is a change to a different brand of coating (paint) than listed in the permit, when the new brand of coating has the same or less VOC content.

The concept of being allowed to contravene a permit with only a seven-day advance notice was vociferously debated. It is clear, however, that, in general, most changes that a source would want to make would not qualify for this category. This would especially be the case if the permit and possible operational scenarios were carefully crafted. The contravening change exemption was included in the regulations primarily to avoid having to reopen permits for minor issues that do not affect emissions.

An important qualifier to all of the above changes is that they must not violate an applicable underlying requirement; they must not increase total emissions over the amount allowed by the permit; and they must not alter the source's monitoring, recordkeeping, or reporting requirements under the permit. These types of changes require permit modifications.

The operational flexibility provisions are summarized in Table 10-3.

TABLE 10-3 Operational Flexibility Provisions for Title V Sources[a]

Provision	Description or Purpose	Requirements or Implications
General provision	Allows source to contravene excessive detail in a Part 70 permit.	Seven-day advance notice is required to contravene permit.
Alternative scenarios	Allows sources to switch to operating modes specified in the permit.	Must be included in application and the permit. Approved ones can be used without notice.
Off-permit operating changes	May allow sources to make certain operational changes not provided for in permit.	Each state may develop specific requirement. Notice required at time of change.
Emissions trading based on the SIP	Allows for emissions trading within plant where authorized by SIP, but permit does not address.	States may provide if the trading is authorized by the SIP. Seven-day advance notice required.
Emissions trading to meet an independent cap	Allows for emissions trading for independent emissions units.	State must provide if an applicant requests a trading cap.

[a]States must develop specific regulations and guidance for these provisions in their operating programs and may have considerable discretion as to how they are applied. None of these provisions can result in a Title I modification, allow a source to exceed its emission limit under the permit, or violate any other applicable requirement.

Permit Modifications

If the changes desired by the source cannot be accommodated within the permit, then the source must seek a permit modification. There are three types of modifications: (1) administrative, (2) minor, and (3) significant.

Administrative modifications are to correct typographical errors and other administrative issues associated with the permit—for example, facility address or change in the name of the certifying official. The permitting agency has 60 days to make these changes; however, they can be implemented immediately after filing with the local agency.

Minor modifications are those that include small emissions increases (less than the *de minimis* levels) or insignificant changes to monitoring or recordkeeping requirements (e.g., changing the frequency of reporting to monthly from quarterly). However, monitoring or recordkeeping changes qualify for a minor modification only if the changes are to make the monitoring and reporting more stringent. The permitting agency has 90 days to make these changes, and the EPA has an additional 45-day period in which to review and veto the changes. During the 135-day review process of both the local agency and the EPA (90 days local plus 45 days EPA), the local authority has the discretion to allow the source to implement the modification. However, there is no permit shield for that modification, and if the agencies deem the modification to be significant or violating an applicable requirement, the source would be subject to enforcement action.

Significant modifications are all modifications other than those that qualify as administrative or minor. This includes significant emissions increases, significant changes in monitoring or recordkeeping requirements, changes in applicable requirements, and so forth. The agency has up to 9 months to make these modifications, and the modifications cannot go into effect until after agency approval.

Permit Reopeners

Even if a source does not wish to reopen its permit in order to make operational changes, the agencies can reopen the permit. The bases for such a reopening are as follows:

1. There are new requirements applicable to the source, and there are more than three years left in the permit term.
2. The source is affected by new acid rain (Title IV) requirements.
3. The state or EPA determines that a permit contains a material mistake or determines that the permit was based on inaccurate information.

4. The state or EPA finds it necessary to reopen and modify the permit to assure compliance with the underlying applicable requirements.

When a permit is reopened, it is handled the same way as an initial permit application, except that only the parts of the permit that are reopened are subject to public review and comment, and the comments and issues must be germane to the cause for reopening. This limitation is to avoid a constant rehash of environmental issues associated with a source.

Public Notice and Review

One of the more troublesome aspects of a Title V permit is that, before it can be issued, there must be ample opportunity for public review and comment. This includes holding a public hearing. Such involvement by the public is common and required prior to issuance of a construction permit for a major new source, but has never been the case for existing sources. There is much concern in industry that for some existing sources the Title V public participation process could become fatal to continued operations. Therefore, many existing sources are evaluating options for avoiding Title V permitting by accepting emissions limits, installing controls, or making process changes such that the emissions are below the Title V major source thresholds.

Another concern related to the public review process is that the EPA also has a formal opportunity and duty to review the permit prior to issuance. Previously, especially for existing sources, the source dealt only with the local permitting authority. Under Title V, however, the EPA will review the permit; if it objects, the local authority must respond to and incorporate EPA comments prior to issuance.

The third area of concern related to public involvement is that neighboring (contiguous) states (where the source could possibly affect air quality in a contiguous state) or states within 50 miles of the source also have the opportunity and duty to review proposed permits. The objections of the neighboring states must also be addressed by the source's local permitting authority.

The fourth area of concern is that, even after all the public, EPA, and neighboring state comments have been addressed by the local permitting authority, any member of the public may petition the EPA within 60 days of the end of EPA's 45-day objection period for additional review and objection. Theoretically, these petitions can only be based on new significant information that was not known or not practicable to be raised during the previous comment periods (i.e., the petition cannot simply restate previous comments). However, traditionally it has been difficult for regulatory agencies to not allow comments from the public. Therefore, it is not certain how these provisions of the Title V permitting program will work out.

Time frames for permit modifications are shown in Table 10-4.

Permit and Application Shields One of the advantages of the Title V permit is that if the applicable requirements are listed in the permit and the permitting authority approves the permit as meeting those requirements, there can be no enforcement action against the facility as long as the facility meets its permit conditions. This is called the *permit shield.* In order for the shield to be most effective, however, the permit must also list those require-

TABLE 10-4 Time Frames for Title V Permit Changes

Type of Review	Applications Due	Time Frame for State/Local Review	Public Review Requirements	EPA Review
Permit issuance	Within a year after EPA approval of state program	3 years to issue all permits[a]	30-day notice and hearing + 60 days to petition EPA	45 days
Permit renewal	At least 6 months before expiration	18 months to complete action[a]	30-day notice and hearing + 60 days to petition EPA	45 days
Permit reopening	States must give affected sources 30 days' notice	18 months after new standard promulgated	30-day notice and hearing + 60 days to petition EPA	45 days
Administrative amendment	When change is implemented	60 days after receipt of application	None	None
Minor permit modification	When change is implemented	90 days after receipt of application[a]	None	45 days
Significant permit modification	When source requests the modification	18 months after receipt of application[a]	30-day notice and hearing + 60 days to petition EPA	45 days
Operational flexibility	7 days before requested action in most cases	Permitting agency must review most changes	None	None

[a]Review by affected states is also required.

ments that do not apply to the source. If those nonapplicable requirements are not listed, the permit shield will not extend to those nonlisted regulations.

There is also an application shield that prohibits the agency from enforcing against a source if the application was filed in a timely manner and has been declared complete and if the source operates in accordance with the application. One common mistake made by sources is not operating in accordance with the application. If this occurs, the application shield will not protect the source from enforcement. It is critical to realize, however, that the permit and application shields extend only to conditions specifically identified, and these conditions cannot violate any underlying applicable requirements.

Annual Emission Fees

It is obvious that the Title V permit program requires a considerable amount of agency time and effort to implement. Therefore, one of the requirements contained in the 1990 CAA was that states must collect emission fees from all Title V sources sufficient to cover the cost of implementing and administering the Title V program on a local level. The amount of those fees can be as large as necessary to completely cover the costs of the state program and can be no less than the amounts necessary. The 1990 CAA establishes a default emission fee value of $25 per ton of emissions per year, in 1990 dollars, with explicit authority to increase the $25 per ton amount by the annual Consumer Price Index (CPI). Therefore, in 1994, when most sources first started paying fees under this program, the default amount was about $29 per ton. The 1990 CAA also established a default maximum emission fee limit of 4000 tons per year per pollutant. If a state incorporates such a maximum value, even if the source emits more than 4000 tons per year, fees can be collected only on the first 4000 tons. Accordingly, if a source emits 3000 tons per year of SO_2, 2000 tons per year of NO_x, and 1000 tons per year of PM_{10}, its annual emission fees would be $174,000 in 1994 dollars, if the local agency used the default CPI-adjusted 1994 value of $29 per ton (i.e., 6000 tons per year of all pollutants times $29 per ton). These emission fees are due on an annual basis as long as the source is a Title V source. Note that individual states have widely varying permit fee programs. For example, some states use allowable instead of actual emissions, some charge fees on all sources (not just Title V), and so forth. Note also that Title V fees are usually collected before a Title V permit is issued.

Annual Compliance Certification

One of the more onerous aspects of a Title V permit is the requirement that a responsible corporate official must annually certify that the source is in compliance with its Title V permit. This certification must be based on information and belief formed after reasonable inquiry that the statements and information are true, accurate, and complete.

A responsible corporate official is defined as a corporate officer, a person in charge of a principal business function, or a plant manager. However, in order for the plant manager or "principal business function person" to be able to sign the certification, the plant manager must have at least 250 employees, $25 million in annual sales, and delegation of authority approved in advance. This degree of certification at such a high level in the corporation is unprecedented in the air quality regulatory arena.

The annual compliance certification must evaluate and certify to each term and condition of the permit, not just the overall emissions amounts. In addition, the certification must indicate whether compliance was continuous or intermittent and what methods were used to establish compliance. Obviously, as stated previously, specifying in detail the compliance methods is an integral element to a good Title V permit.

Upsets, Breakdowns, and Malfunctions

Whenever mechanical equipment is operated, there will be upsets, breakdowns, and malfunctions. The question arises, How are these upsets reported in light of the annual certification? This issue is addressed in two ways: (1) Upset conditions should be identified and included in the permit operational scenarios, and (2) true emergency upsets can be reported to the agency and enforcement barred.

In the past, permits for sources were generally issued for maximum potential emissions under full capacity and with controls installed and operating at their normal removal efficiency. Therefore, in the past, control system malfunctions were not covered in the permit. However, the facility was protected from enforcement as long as the malfunction was reported to the agency in accordance with the agency's upset breakdown rules. In order to have similar protection under Title V, however, it will not be sufficient to avoid listing malfunction conditions in the permit and then report the upsets. If the upset conditions are not listed in the Title V permit, the annual certification of compliance will not be able to be provided, and the permit shield will not apply. Therefore, as part of the operational scenarios, reasonably anticipated upset provisions should be evaluated and included in the permit. This may be a difficult concept to get through the local permitting authority, but some protection for upsets needs to be included in the Title V permit.

The second method of protection from upsets causing noncompliance is to promptly report the upset or malfunction to the permitting authority as an emergency condition. In order to qualify as an emergency condition, it must meet all of the following tests:

1. The condition must be a reasonably unforeseeable event beyond the control of the source.
2. The event must have required immediate corrective action of the source in order to restore normal operation.
3. The event must *not* be due to poorly designed equipment, lack of preventive maintenance, careless or improper operation, or operator error (these are the same types of conditions used in the Clean Water Act NPDES permits).
4. All reasonable steps were taken in order to minimize emissions and comply with the permit.
5. The permitting authority was notified in writing within two working days of the emergency.

In the past, some sources in some jurisdictions have been allowed to file literally hundreds of upset or breakdown reports. Such practice is not likely to be allowed under a Title V permit, especially in light of the public review process. Therefore, it is critical to ensure that the permit allows for malfunctions and that equipment is maintained to perform at the level required by the permit.

EFFECTIVE TITLE V PERMITTING

In order to prepare and submit an effective Title V permit or permit modification, there are several steps that should be taken:

1. *Identify the full range of operational needs* that the facility may have, even if those needs are not permitted or even permittable. This extensive list of needs will then be pared down and summarized into feasible operational scenarios. If, however, one does not identify the full range of needs, there is a high probability of missing a scenario and needing a permit modification later.

2. *Identify all of the applicable underlying regulatory requirements* and those regulatory requirements not applicable to the facility and emitting devices at the facility. Identify future possible requirements and determine

how those requirements could be met. Although it may not be necessary to incorporate future requirements into the permit at the time of application, if one anticipates the requirements, a facility may be able to implement the requirements early and avoid Title V or a reopener.

3. *Determine monitoring and compliance methods and techniques.* Determine what methods of compliance testing are feasible, what methods are required by the local permitting authority, what methods are amenable to routine recordkeeping at the facility, and what methods are infeasible from a practical perspective. Because the facility is required to submit an annual compliance certification, it is very important that the chosen methods of compliance are simple and feasible to consistently implement on a daily basis.

4. *Determine the actual emissions and the maximum potential to emit* of any regulated pollutant. Identify and document *de minimis* emissions and insignificant activities. Carefully document the basis for the emission estimates. Review the emissions factors for representativeness, including those published by the regulatory agencies (such as the EPA publication AP-42) and those provided by equipment vendors. Ensure that the emission factors and techniques of estimating emissions are representative of the specific installation. Factors to be considered include the age of the equipment, how it is maintained, and how it is used.

5. *Develop operational scenarios.* Test the scenarios and refine them to minimize their number and complexity. The more complex and limiting the scenarios, the harder it will be to comply with the scenarios.

6. *Evaluate recordkeeping and monitoring methods* and select those methods that will successfully demonstrate compliance with chosen operational scenarios.

7. *Refine and reduce the scenarios* consistent with the monitoring and recordkeeping practices that the facility can accept on a routine basis.

8. *Prepare draft permit application and permit conditions.* Submit draft conditions and monitoring techniques to the permitting authority, even if the agency does not explicitly ask for such information. As mentioned in Chapter 2, the agency does not have the time or specific knowledge of your specific facility to accurately prepare scenarios and permit conditions.

The EPA has published several model Title V permits (termed "hollow permits") that provide a format and some indication of the types of data that will be needed in the permit application. The hollow permits also indicate how compliance will be determined. Appendices N and O contain copies of two types of hollow permits: one for PM_{10} emissions from lime manufacturing plants and one for graphic arts facilities. As of late 1995, the EPA had published draft hollow permits for about 35 different facilities. Among the published hollow permits are the following types of sources:

- PM_{10} emissions from lime manufacturing plants
- Light duty truck coating operations
- Coating of flat wood paneling
- Bulk gasoline plants
- Bulk gasoline terminals
- Gasoline dispensing facilities—Stage I vapor recovery
- Gasoline tank trucks
- Solvent metal cleaning
- Petroleum refinery sources (other than leaks, tanks, and terminals)
- Petroleum liquid storage in external floating roof tanks
- Petroleum liquid storage in fixed roof tanks
- Cutback and emulsified asphalt
- Synthesized pharmaceutical products
- Pneumatic rubber tire manufacturing
- Petroleum solvent dry cleaners
- High density polyethylene, polypropylene, and polystyrene resins
- Synthetic organic chemical manufacturing industry
- Graphic arts
- Lead acid battery manufacturing
- Secondary lead smelting
- Secondary lead remelting
- Iron and steel mills
- Nonferrous smelters
- Primary aluminum smelters
- Pulp mills
- Utility boilers
- Paint manufacturing
- Elemental phosphorus production
- Glass manufacturing
- Portland cement manufacturing
- Marine grain elevators
- Grain milling
- Nonmetallic mineral processing
- NESHAP permits for mercury, arsenic, polyvinylchloride, ethylene dichloride, vinyl chloride

SUMMARY

After a Title V permit application is submitted, careful coordination with the agency will be needed to ensure that the issued permit is consistent with actual operations. When the permit is issued, diligent monitoring and record-keeping will be necessary to ensure that the facility operates in accordance with the permit. If this type of diligence is not exercised, the enforcement penalties discussed in Chapter 13 can become a serious reality.

CHAPTER

11

ACID RAIN PROVISIONS OF THE 1990 CAA

Although the acid rain provisions (i.e., Title IV) of the 1990 CAA affect only a relatively small number of sources, the air quality improvement expected from Title IV is significant. In addition, Title IV implemented two innovative concepts, nationwide emissions trading and a national emissions cap, that are likely to be implemented in the future on a much broader scale.

In summary, Title IV reduces total nationwide electric utility emissions of SO_2 by 10 million tons per year by the year 2000, a 50 percent reduction from emissions in 1980 (1980 utility emissions totaled about 20 million tons per year). Furthermore, Title IV caps the total electric utility SO_2 emissions at the 10-million-ton-per-year level (i.e., no increases in total SO_2 emissions can occur in the future). The emissions cap has been referred to as a "green field cap." Title IV accomplishes these goals by establishing a nationwide system of SO_2 emission allowances. Emission allowances are in terms of tons of SO_2 emissions per year and are an authorization to emit 1 ton per year of SO_2. These allowances are similar to the ERCs (emission reduction credits) used in the nonattainment program to offset increases in emissions. No source affected by Title IV can emit SO_2 without holding SO_2 emissions allowances. If a source wishes to increase production (and, therefore, emissions), it must first obtain emission allowances from some other source that has reduced emissions.

In addition to the SO_2 reductions, Title IV requires a 2-million-ton reduction in NO_x emissions. This reduction is to be accomplished through burner modifications (i.e., Title IV does not require installation of postcombustion controls, such as selective catalytic reduction, to reduce NO_x emissions). The 2-million-ton-per-year reduction is measured from the level of emissions that would have occurred in the year 2000 without installation of low-NO_x burners. There is no NO_x emission allowance system or emissions cap.

AFFECTED SOURCES

Title IV applies only to "affected sources." These are emission sources that are affected by Title IV, which include utility steam electric generating units that generate more than 25 megawatts of electricity (MWe) and other smaller generating units or industrial sources that choose to become affected, termed "opt-in" sources.

There are four groups of affected sources:

1. *Phase I Sources.* Phase I sources are the 110 electric generating plants (consisting of 261 individual units) that were existing on November 15, 1990 (when the 1990 CAA was signed) for which the Act (not regulations) established specific limits on emissions (in terms of tons per year). The Phase I sources were those units that emitted relatively large amounts of SO_2. Due to the high-sulfur coal located in the eastern United States, these Phase I sources were all located east of the Mississippi River.
2. *Phase II Sources.* This category of sources includes all (existing and new) electric utility steam electric generating units larger than 25 MWe (including the Phase I sources). It is estimated that about 2200 individual units are affected by Phase II. Phase II units do not include *existing* combined-cycle plants or existing qualifying small power production or qualifying cogeneration facilities under the Public Utility Regulatory Policies Act (PURPA). However, new (i.e., post-1990 combined-cycle and cogeneration facilities) are included in Phase II.
3. *New Units.* These are units that commenced operation after the 1990 CAA was enacted.
4. *Opt-in Sources.* These are sources not affected by Title IV, but that choose to become affected by the Title so that they can sell their SO_2 emission allowances to electric utilities.

EMISSION ALLOWANCES

The fundamental concept of Title IV is that emitting sources will be given (or required to obtain) emission allowances that allow the source to emit SO_2. These allowances are similar to the emission reduction credits (ERCs) that must be obtained by new or existing sources to offset emissions increases in nonattainment areas. One of the main differences, however, is that SO_2 allowances must be held for *all* emissions at *existing or new* sources under Title IV, whereas ERCs must be held only for new and modified existing sources in nonattainment areas and only for the incremental increase. A second difference is that emission allowances are issued every year on an annual basis, whereas once ERCs are used to offset emission increases (from new or modified sources), the ERCs are surrendered.

Phase I Emission Limits

Phase I sources were defined and limited in order to accomplish SO_2 emissions reductions of about 5 million tons per year from the 1980 baseline. In other words, in 1980 these 110 sources emitted about 10 million tons per year of SO_2 and, after meeting the legislated Phase I source limits, the total from those 110 plants will be only about 5 million tons. Phase I sources must meet the legislated limits no later than January 1, 1995, or obtain an extension from the EPA along with additional allowances for the excess emissions. If the Phase I sources do not meet the legislated limits or obtain the extensions and allowances, they will face severe penalties.

The emission allowances for the Phase I sources were established based on an SO_2 emission rate of 2.5 pounds of SO_2 per million Btu heat input (i.e., 2.5 lb/mmBtu). (Note that in 1980 some of the Phase I units emitted more than 12 lb/mmBtu.) However, the emissions limit is not in terms of the rate (i.e., pounds per mmBtu), but rather in terms of an annual emissions quantity (tons per year). Therefore, the sources can meet the Phase I limits by fuel switching (e.g., switching from high-sulfur to low-sulfur coal, or switching to natural gas), installing flue gas desulfurization (FGD) controls (e.g., SO_2 scrubbers), shifting electric production from dirty to clean plants, encouraging more efficient electric use by consumers (through conservation), obtaining other allowances as described above, or any combination of techniques. This Title IV concept of flexible options to meet emission limits is also embodied into the Title V permit program and is currently being incorporated into several local permitting programs (e.g., the RECLAIM program of Los Angeles; see Chapter

9). Note that Phase I reductions do not result in emission reduction credits (Chapter 9) because the reductions are not "surplus" in that the reductions were required by rule.

If a Phase I source reduced emissions to levels below 1.2 lb/mmBtu prior to the 1995 deadline, the source could receive two-for-one allowances for the reduction. In other words, if a source reduced to below 1.2 lb/mmBtu, the amount of reduction times two could be used at other plants to meet the Phase I limits. In addition, if a unit is retired, the allowances for that unit can be used at other locations in the utility system. Finally, if a utility employs renewable energy technology (e.g., solar, geothermal, biomass, or wind) resources for some portion of its generation mix, the EPA can issue bonus allowances to the utility.

In summary, Phase I units have two general techniques to meet the Phase I limits: Adjust operations at the unit to meet the limit, or transfer allowances to the unit. The operational adjustments include fuel switching, reduced operations, or FGD controls. Transfer of allowances are from some other retired unit, some other controlled unit, or some other repowered unit (with renewable energy resources).

Phase II Emission Limits

All Phase II sources must meet emission limits calculated from an emission rate of 1.2 lb/mmBtu, and those limits must be met by January 1, 2000. In addition, if the unit emits less than 1.2 lb/mmBtu in 1985, the allowances are set at the actual 1985 emission rates (i.e., those units will not be allowed to increase emissions above 1985 baseline values). If the emission limits are not met in Phase II, additional emission allowances must be obtained. However, there is no prohibition for an individual unit emitting more than 1.2 lb/mmBtu, as long as sufficient allowances are available.

The calculation of Phase II allowances is fairly complicated, because units are given special allowances to account for special operating conditions that a unit may have had in 1985. These special conditions include allowances for units that burned primarily natural gas in 1985, but are also capable of burning oil; allowances for units that began operation after 1985; allowances for units operating at unusually low capacity factors in 1985; and allowances for units in smaller utility systems, units in Florida, units owned by municipalities, and units in certain states with historically low emission rates. Most of the special allowances are designed to provide allowances to high growth states and states that have had historically very low emissions due to low sulfur coal usage, high degree of FGD control required, or both.

The EPA has completed calculation of allowances accounting for the above factors and has published a list of units and allowances. Some of these allowances have already been traded on the open market.

NEW SOURCES

No new unit can operate without obtaining allowances. These allowances can be from existing units that are retired or controlled so that the existing unit's allowances can be transferred to the new unit or from the EPA through allowance sale or auctions.

New sources owned and operated by an independent power producer (IPP) must also obtain allowances prior to operation. However, the Act provides that the IPPs be guaranteed the right to purchase allowances from the EPA at a default rate of $1500 a ton (adjusted for inflation) or at auction. This allows IPPs to be certain that allowances will be available, because, contrary to electric utilities with existing units, there are no other sources of allowances for IPPs. (It is assumed that electric utilities may not want to sell their allowances to an IPP, because the IPP would essentially be a direct competitor to the utility.)

OPT-IN SOURCES

If a source that is not affected by Title IV chooses to become an affected source, the emissions of that source become emission allowances. These allowances can then be transferred to other affected sources. The amount of allowances are the actual or permitted emissions of the unit, whichever is less, in 1985 based on the 1985 emission rate and average operational data (e.g., fuel use) for the period from 1985 to 1987. If the allowable current emission rate (lb/mmBtu) at the time of the application (not total emissions) is less than the rate in 1985, the EPA will use the lower emission rate and apply it to the 1985 to 1987 baseline operational data. (This prohibits double credit for emissions reductions that have occurred since 1985.)

Any source that is exempt *by regulation* from Title IV (e.g., utility boilers with less than 25-MWe generation, burning fuel with less than 0.05 percent by weight sulfur, or exempted retired units per 40 CFR 72.8) *cannot* opt-in to Title IV.

The advantage of opting-in to the program is the ability to sell emission allowances. There are two major disadvantages, however. One is that a Title IV source is automatically a Title V source, and a previously exempt source would, if it opted-in to Title IV, have to obtain a Title V operating permit. Secondly, once a source opts-in to Title IV, emission allowances will be established for that source and the source cannot increase emissions above its allowances without obtaining additional allowances elsewhere.

It is generally assumed that if electric utilities need allowances to provide for future expansion and growth, they may contact industrial sources, convince those sources to opt-in to Title IV, replace antiquated devices that burn high-sulfur fuels with electric or other low-emitting devices, and then transfer the allowances to the new electric generating units. If the opt-in unit achieves emission reductions through replacement of devices or emissions control technology, the allowances given to the opt-in unit may be transferred to any other Title IV source. For example, if the opt-in unit installs an emissions control device, the reductions can be transferred to other Title IV sources.

If, however, emission reductions at the opt-in unit occur through shutdown or reduced utilization of a combustion device, the allowances generated through this means can be transferred to other affected units only to the extent that the reduced utilization or shutdown resulted from transferring thermal energy from some other affected unit to the opt-in source. This restriction is designed to prevent electric utility companies from transferring allowances to an opt-in unit and then shutting down that unit to use the allowances elsewhere. That is, this restriction prevents a "shell game" within the Title IV universe of sources. However, if the shell game is, for example, designed to shift thermal energy production from an affected unit to an opt-in unit, then that amount of thermal energy shifted is allowed to generate transferable allowances. It remains to be seen how many sources choose to opt-in to the Title IV program and how many allowance trades occur through this vehicle.

PERMITS AND COMPLIANCE PLANS

Title V Operating Permits

All Title IV sources (whether they are affected sources or opt-in sources) must have a Title V national operating permit. For the Phase I sources, the permit comes from the EPA and was required to be issued by the end of 1993. For Phase II sources, however, and at the expiration of the initial Phase I source permits (i.e., 1998), permits will be issued by the local permitting authority as a Title V national operating permit. (The special system for Phase I sources was necessary because the states' Title V program was not required to be implemented until 1994/1995.)

Compliance Plans

Title IV requires a source to submit compliance plans as to how they will achieve its emissions allowances and how compliance with the allowances will be determined. Under Title IV, the EPA has indicated that these compliance plans can be very straightforward and simply written. This is to allow the greatest flexibility. However, when the Title V operating permits are issued for the Title IV sources, the compliance plans may have to become much more complex.

Monitoring, Recordkeeping, and Reporting

All Title IV sources must have installed continuous emissions monitoring equipment (CEMs). Phase I sources must have installed it by November 15, 1993, and Phase II sources by January 1, 1995. New sources must have CEMs installed before commencement of commercial operation. Opt-in sources must have CEMs installed before they become Title IV sources (i.e., their application for opt-in has been approved and emission allowances authorized by the EPA). The monitoring and recordkeeping aspect of Title IV is extremely complex due to monitoring equipment precision and reliability, the need to calculate allowances in terms of tons per year, equipment downtime, and so forth.

EXCESS EMISSIONS PENALTIES

If a Title IV source exceeds its annual emissions allowance, the source must pay a penalty to the EPA equal to $2000 per ton of emissions that exceeds the allowance. Note that allowances are on an annual basis. Therefore, a source is allowed to exceed shorter-term emission levels (e.g., tons per day calculated by dividing the annual allowances in tons per year by 365 days per year) as long as the total emissions for the year are less than the allowances. Note that if a source exceeds its emission allowances, there may also be a Title V operating permit violation. Similarly, if the Title V permit is not correctly written, or more restrictive than Title IV requires, there may also be a requirement to meet shorter-term emission limits (e.g., pounds per day). In addition to the penalty, the source's allowances for the year following the exceedance must be reduced by the amount of excess in the preceding year.

NO_X REDUCTIONS

Although the SO_2 emission allowance system of Title IV has received the most attention, as it is believed that nitrogen oxides (NO_x) also contribute to acid deposition, Title IV includes provisions to restrict the emissions of NO_x from electric utilities. However, contrary to the SO_2 provisions, the NO_x provisions are structured along the lines of the classic "command and control" system.

The Title IV acid deposition provisions are designed to achieve a 2-million-ton-per-year reduction in NO_x emissions from affected units. The reduction is measured from that level of NO_x emissions projected for the year 2000. This is to be achieved by EPA promulgating emissions limits on affected sources. The emission limits must be no greater than 0.45 lb/mmBtu for tangentially fired boilers and 0.50 lb/mmBtu for wall-fired dry-bottom boilers. These limits were to be met by January 1, 1995, and proposed regulations to meet these requirements have been published. By January 1, 1997, the EPA is to promulgate emission limits on a lb/mmBtu basis for wet-bottom wall-fired boilers, cyclones, cell burners, and all other types of utility boilers. However, none of the emissions limits to be promulgated can require more than burner modifications (termed "low-NO_x" burners). If low-NO_x burners can yield emission limits less than the 0.45 and 0.50 lb/mmBtu level, then the EPA has the authority to promulgate the more stringent emission limits.

The NO_x provisions of Title IV only apply to affected sources. Because there is no emissions trading provision associated with the NO_x provisions, there are no opt-in sources, and there is no flexibility available to the electric utilities to meet the NO_x limitations.

SUMMARY

The SO_2 provisions of Title IV embody two new concepts: nationwide emissions trading and a national emissions cap. It will achieve something that environmental groups have been attempting since the beginning of the movement: a maximum amount of emissions that cannot increase, regardless of future population and industrial growth. It also establishes an innovative market-based emissions control system that is supposed to minimize the cost of compliance. It is these aspects of Title IV that are often discussed as being needed at the local level and that some jurisdictions have already implemented to a significant degree.

CHAPTER 12

STRATOSPHERIC OZONE PROTECTION

Title VI of the 1990 CAA pertains to stratospheric ozone protection. As has been previously stated, in the lower atmosphere (termed the *troposphere*), there is too much ozone in more than 100 cities in the United States (i.e., the NAAQS for ozone is exceeded). However, in the upper atmosphere (termed the *stratosphere*), it is believed that certain substances consume ozone through high-energy photochemical reactions. These chemicals are referred to as *ozone-depleting chemicals* (ODCs) or *ozone-depleting substances* (ODSs).

ODCs are chemicals that are relatively stable in the lower atmosphere (such as freons), where the sun's energy has been attenuated by the air, but decompose when exposed to high-energy light from the sun in the upper atmosphere. In addition, when ODCs do decompose in the atmosphere, the decomposition process results in free radicals that readily react with ozone, thereby reducing the amount of ozone in the upper atmosphere. On the other hand, VOCs that are regulated in order to prevent formation of ozone in the lower atmosphere are chemicals that rapidly undergo decomposition in the lower atmosphere and react with hydrocarbons and other chemicals in the air to form ozone.

The distinction between a chemical's ability to react in the lower and upper atmospheres can produce seemingly contradictory regulatory requirements. For example, 1,1,1-trichloroethane (TCA or methyl chloroform) is often substituted for other VOCs, because TCA does not react in the lower atmosphere to form ozone. Unfortunately, however, TCA reacts readily in the upper atmosphere and is an ODC.

TITLE VI REGULATORY PROGRAMS

Contrary to many of the 1990 CAA programs, the stratospheric ozone protection program is relatively straightforward and is implemented automatically in accordance with preestablished dates. Many of the provisions of Title VI do not require promulgation of implementing regulations, and none of the provisions requires states to incorporate Title VI regulations in their SIPs. This is due to the fact that SIPs are designed to comply with the NAAQS (which are not applicable to ODCs) and the fact that Title VI is designed to comply with the Montreal Protocol to phase out ODCs. The Montreal Protocol was signed and enacted by Congress and has its own requirements. Therefore, additional new regulations are often not necessary.

The Title VI provisions can be summarized as follows:

1. Phaseout of production of ODCs
2. Labeling requirement for products manufactured with ODCs
3. Ban on production of nonessential products containing ODCs
4. Work practice standards to prevent release of ODCs

PHASEOUT OF PRODUCTION OF ODCS

Chemical Lists

The CAA requires the EPA to list two classes of ODCs, depending on the ozone depleting potential (ODP) of the chemical. (The ODP is a relative measure of the ability of a chemical to consume ozone in the upper atmosphere. High ODPs are in the range of 1 and low ODPs are in the range of 0.1 or less. The ODP is referenced to the ozone deleting potential of CFC-11, freon.) The 1990 CAA established a minimum list of ODCs for the two classes, called Class I and Class II chemicals. However, in response to petitions or on its own accord, the EPA can add to the list (but not delete chemicals from the list).

Within the two classes of ODCs, the Act establishes various groups of chemicals. The chemicals are grouped in order to allow for chemical substitutions within a group as the manufacturers of the chemical strive toward the phaseout deadlines.

The current list of Class I ODCs is as follows:

Group I: Chlorofluorocarbon (CFC)-11, 12, 113, 114, and 115

Group II: Halon-1211, 1301, 2402

Group III: CFC-13, 111, 112, 211, 212, 213, 214, 215, 216, and 217

Group IV: Carbon tetrachloride

Group V: Methyl chloroform (TCA)

Group VI: Methyl bromide

Group VII: Hydrobromofluorocarbons (HBFC)

The current list of Class II substances is as follows:

Hydrochlorofluorocarbon (HCFC)-21, 22, 31, 121, 122, 123, 124, 131, 132, 133, 141, 142, 221, 222, 223, 224, 225, 226, 231, 232, 233, 234, 235, 241, 242, 243, 244, 251, 252, 253, 261, 262, and 271

Phaseout

The Class I and II ODCs must be phased out of production in accordance with the Montreal Protocol, the 1990 CAA, and the EPA's own plans. The CAA established a phaseout schedule that would have resulted in all Class I ODCs being no longer produced by January 1, 2000 for all chemicals except TCA and methyl bromide. However, the EPA accelerated the phaseout schedule so that halons were eliminated by January 1, 1994 and so that the CFCs, carbon tetrachloride, TCA, and HBFC were eliminated from production by January 1, 1996. The EPA has ordered a complete elimination of methyl bromide by January 1, 2001, and a freeze on production to no more than 1991 levels, beginning January 1, 1994. In addition to the complete phaseout dates, the EPA has also established interim phaseout quantities as shown in Table 12-1. Phaseout of Class II ODCs has been proposed by the EPA to follow the schedule as shown in Table 12-2.

The Act provides, for specific ODCs for specific amounts of time, some limited exemptions to the phaseout schedule. These exemptions are related to necessary uses for medical devices, aviation safety, national security, fire protection, and certain uses of TCA for which no safe substitute is available.

■ TABLE 12-1 Class I ODC Phaseout Schedule (As of July 1, 1994) (40 CFR 82.6)

Year Beginning January 1	Allowed Percentages[a]					
	CFCs	Halons	Carbon Tet	TCA	Methyl Bromide	HBFC
1994	25	0	50	50	100	100
1995	25	0	15	30	100	100
1996	0	0	0	0	100	0
1997	0	0	0	0	100	0
1998	0	0	0	0	100	0
1999	0	0	0	0	100	0
2000	0	0	0	0	100	0
2001	0	0	0	0	0	0

[a]Percentages of baseline 1993 production allowed in 1994–2001.

■ TABLE 12-2 Class II ODC Phaseout Schedule (Proposed Mar. 13, 1993) (58 FR 15014)

Year Beginning January 1	ODC	Restriction
2003	HCFC-1416	Ban on consumption and production.
2010	HCFC-22 HCFC-142b	Production and consumption frozen at baseline levels. Ban on use of virgin chemical except as feedstock or refrigerant in appliances manufactured prior to January 1, 2010.
2015	All other HCFCs	Same restrictions as 2010 limits.
2020	HCFC-22 HCFC-142b	Ban on consumption and production.
2030	All other HCFCs	Ban on consumption and production.

LABELING PRODUCTS MANUFACTURED WITH ODC

Section 611 of the Act requires any company that manufactures products with Class I ODCs to label, beginning May 15, 1993, the manufactured product with the following statement:

> Warning: Manufactured with (insert name of ODC), a substance which harms public health and environment by destroying ozone in the upper atmosphere.

A similar warning label is required on containers and products containing Class I ODCs.

Definition of "Manufactured With"

The term "manufactured with" means that the manufacturer uses the ODC directly in the product's manufacturing process, but the product does not contain the ODC. If the product *contained* an ODC, it would have to follow the container labeling requirement. In order for a product to have been "manufactured with" ODCs, it must have been in physical contact with the ODC on more than an intermittent basis. (Note the distinction between "manufactured with" and "contained in.")

If a second manufacturer uses a product that was manufactured by a first manufacturer with ODCs, the pass-through exemption applies. This means that the EPA does not require the label to remain on a product if it is passed through to the consumer in the second product. This pass-through exemption means that the second manufacturer can remove the label on a product that was originally made with ODCs if the following applies to the second manufacturer:

1. The product was purchased from another manufacturer (subsidiaries and other divisions of the same company do not qualify for the pass-through exemption even if the separate division is located overseas).
2. The purchased product does not *contain* ODCs.
3. The manufacturer does not use ODCs in its own manufacturing process.

Interstate Commerce Exemption

The warning requirements apply only to products that enter into interstate commerce. (Therefore, if a product is made only for export, the labeling requirement would not apply.)

De minimis Exemptions

The container labeling requirement is not required if the ODC is present only as a trace constituent, residue, or impurity, and the ODC serves no useful purpose in the product contained in the container.

The "manufactured with" labeling requirement does not apply when the product manufactured only comes into contact with the ODC in an intermittent manner, not as a routine part of the manufacturing process. A second *de minimis* exemption occurs when the manufacturer uses CFC-13 or TCA as solvents, and the company undertook a program that reduced usage of TCA or CFC-13 by 95 percent from 1990 levels by May 15, 1994.

Other Exemptions

A product is not manufactured with ODCs if the ODC has been transformed, except for trace quantities, before the ODC came into physical contact with the manufactured product.

A product is not manufactured with ODCs if the product came into contact with substances that contained an ODC, if the ODC is generated through a coincidental unavoidable by-product (CUBP). In order to qualify as a CUBP, the ODC so generated must have been immediately captured and destroyed to at least 99.99 percent efficiency.

ODCs used in repairs or maintenance of manufacturing equipment do not qualify as being "manufactured with." Likewise, products made as part of a research and development effort do not require labeling until the products enter into interstate commerce.

A manufacturer can petition the EPA for an exemption to the labeling requirement. However, the exemption must be product-specific, and it must be demonstrated that there are no substitute products or manufacturing processes that (a) do not rely on the use of such ODC, (b) reduce the overall risks to human health and the environment, and (c) are currently or potentially available. No such exemption is available for the container labeling requirement.

BAN ON NONESSENTIAL PRODUCTS

Nonessential products containing or manufactured with Class I ODCs were banned effective February 16, 1993, except for sales to commercial purchasers, who had until January 1, 1994. Such products containing Class II ODCs were banned effective January 1, 1994, except in the case of integral skin foam utilized to provide vehicle safety, which can continue to January 1, 1996. The nonessential products include (as general categories):

- String confetti
- Noise horns
- Cleaning fluid
- Plastic flexible packaging material
- Aerosol products

(Note: There are several specific exemptions and types of products that may be used in certain applications—for example, medical devices. Therefore, the Federal Regulations, 40 CFR Part 82, Subpart C, need to be consulted for details.)

WORK PRACTICES

In addition to banning production of ODCs and products manufactured with ODCs, the EPA recognizes that there are still large quantities of ODCs used by the public. Therefore, the Agency has promulgated regulations that prohibit the release of ODCs during the practice of maintaining, servicing, repairing, or disposing of an appliance or industrial process refrigeration. Automobile air conditioning systems must be serviced only by EPA-trained and certified technicians in order to avoid having ODCs being released to the atmosphere. In addition, effective November 15, 1992, the sale of small containers of CFCs (for do-it-yourself automobile air conditioner servicing) was banned except to certified technicians.

IMPLICATIONS

As stated, one of the interesting Class I ODCs is TCA. This chemical has been widely and routinely "forced" on industry as a substitute for other VOCs (e.g., perchloroethylene) as a degreasing solvent and as an auxiliary blowing agent for the manufacture of polyurethane foam. The quantities of TCA used are very large (on the order of 1,000,000 pounds per year for a typical midsized foam manufacturing plant). The ODC phaseout schedule has placed these industries in a precarious position, because one agency (the local permitting authority) is forcing them to use TCA while the other agency (EPA) is forcing phaseout of the chemical. This type of double bind is not uncommon in environmental regulations and emphasizes the need for multimedia regulatory approaches where all aspects of an environmental issue are evaluated before promulgating individual regulations.

CHAPTER 13

ENFORCEMENT

The 1990 CAA has greatly increased the enforcement and penalty provisions related to potential air quality noncompliance. Some observers have concluded that the penalties related to air quality are now the most stringent of any of the federal environmental laws. However, as other environmental laws are reauthorized [e.g., the Resource Conservation and Recovery Act (RCRA) and the Comprehensive Environmental Response, Cleanup, and Liability Act (CERCLA) or "Superfund"], those acts will probably contain even harsher penalties.

The 1990 CAA enforcement provisions contain significant civil and criminal penalties. The provisions embody the following concepts:

1. Criminal recordkeeping violations
2. Criminal penalties for knowing endangerment
3. Criminal penalties for negligent endangerment
4. "Continuous" civil penalties
5. Administrative penalties without filing suit
6. Administrative orders for as long as one year

7. Field citations
8. Banned contractor listings for facilities and owners of facilities that violate the Act
9. Bounty hunter provisions
10. Citizen suits to enforce compliance.

Each of the above concepts will be discussed in the following sections. In addition, when the provision is different from similar provisions incorporated into previous Acts (e.g., the 1977 CAA), the difference will be noted. Although the following sections explain the concepts behind the various enforcement penalties, should a facility be faced with such penalties, the advice of competent legal counsel is necessary. The penalties are potentially very severe ($25,000 per day per violation for civil penalties and up to $1,000,000 or more plus 7 years in jail for criminal penalties). In addition, the legal interpretations, judicial interpretations, legislative history, Agency policies, and other information greatly affect how large and what type of penalties will be required. Therefore, when faced with an actual or potential violation, legal counsel should be obtained.

The information in the following sections is based in large part on an article by Elliott et al., published in the *Journal of the Air and Waste Management Association* (Elliott et al., 1992).

DEFINITION OF VIOLATION

Most of the enforcement penalties are assessed "per violation." This is defined to include each day of violation for each pollutant that is the subject of the violation. As an example, assume that a large facility has a wet sulfur dioxide scrubber as part of its postcombustion control system and that the scrubber is required to be operational as part of the permit. If the operator of the facility shuts down the scrubber, it is likely that not only will the sulfur dioxide permit limit be exceeded, but the particulate limit as well. In addition, if the particulate limit is exceeded, the opacity limit is also likely to be exceeded. Therefore, for each day that the scrubber is off-line without permission of the Agency, there will be at least three violations: sulfur dioxide, particulate, and opacity. If the violations continue for 30 days and the penalty is $25,000 per day per violation, the total potential penalty would be 3 times 30 times $25,000 or $2,250,000! When it comes to enforcing penalties, the Act requires the Agency to presume that the violations occur continuously.

FEDERAL, STATE, AND LOCAL ENFORCEMENT

The following sections discuss federal enforcement provisions only. However, the local permitting authorities also have local enforcement capability and penalties. (In fact, the Act requires states and local permitting authorities to have powers to enforce Title V permits; see Chapter 10). In the case in which a facility has a federally enforceable limitation in a permit obtained from a local permitting authority, the facility could be subject to both local and federal penalties.

The degree of federal enforceability has been increased in the 1990 CAA. Therefore, it is no longer the case as in the past, when a facility could work out an enforcement issue with the local permitting authority and rarely, if ever, be subject to additional penalties from the federal level. This can and has recently become a serious problem in cases in which a facility obtained a formal or informal variance from the local permitting authority but the EPA did not recognize the variance as being valid. Although these cases are usually defensible, they do cost additional hardship and money to resolve. Therefore, a facility has to be diligent in attempting to ensure that EPA supports local decisions.

CRIMINAL RECORDKEEPING VIOLATIONS

The Act now makes it a criminal offense (felony) for an individual or corporation to fail to keep, maintain, or report material information related to compliance. Such information includes documents, required monitoring equipment, making false statements or certifications, and failure to report noncompliance as required by law.

The penalties include fines up to $250,000 for an individual plus two years' imprisonment for the first offense and double those amounts for a second offense. Corporations can be fined up to $500,000.

CRIMINAL KNOWING ENDANGERMENT

This provision is the most severe and includes felony offenses for knowingly releasing imminently dangerous amounts of hazardous air pollutants listed in the Clean Air Act (Chapter 7) or Extremely Hazardous Substances (EHSs) listed in the Emergency Planning and Community Right-to-Know Act (EPCRA). In order for these provisions to apply, the alleged violator must have known that the releases would place others in imminent danger of serious bodily harm or death.

Punishment under these provisions includes a felony conviction and individual fines up to $250,000 and 15 years in jail, while corporations are subject to fines up to $1,000,000 for each violation. Again, second offenses are double these maximum amounts.

CRIMINAL NEGLIGENT ENDANGERMENT

A more common violation would be negligent endangerment. In this case, it is not necessary to prove that the alleged violator knew that the emissions posed an imminent danger to the public, but only that the alleged violator allowed the emissions to occur. These provisions apply to the same list of pollutants as knowing endangerment (HAPs and EHSs). However, it is a defense to these provisions if the releases were in accordance with a permit or if the person endangered consented to being so exposed.

Penalties for this offense are a misdemeanor charge and fines up to $250,000 and one year in jail, and double for a second offense.

CONTINUOUS CIVIL VIOLATIONS

When a violation is discovered, the Act states that it will be presumed that the violation occurred continuously from the time of the occurrence (*not* the time of discovery) until the violator can *prove* that the violation ceased. Therefore, the violator has the burden of proving that there were intervening days when the violation did not occur (e.g., on some days a facility may have been operating at reduced capacity, such that even if the control device was not fully operational, the emissions from the device would still be below the permitted emission limit). The violator also has the burden of proof to demonstrate that the violation has been remedied. For example, it is no longer sufficient to simply repair the control device, but now the EPA may require the facility to conduct a source emissions test to demonstrate compliance. The period of violation in this case could extend until the time that the Agency receives the source emissions test report.

The penalty under this type of provision is a civil penalty, assessed against the corporation in the amount up to $25,000 per day per violation.

ADMINISTRATIVE PENALTIES

One of the new civil enforcement mechanisms incorporated into the 1990 CAA is to allow the EPA to issue administrative penalties of up to $25,000 per day per violation up to a maximum of $200,000 *without* the necessity to file a lawsuit in civil court. Previously, if the Agency wished to fine a facility, it had to file a lawsuit through the Department of Justice. Under the 1990 CAA, the appeal of such administrative penalties is to the EPA Administrative Appeals Board. If the alleged violator fails before the Appeals Board, the decision can be appealed to federal district court, but the judge in that case will be reviewing the body of evidence presented before the Appeals Board, not obtaining new testimony or facts. Therefore, in general, it will probably be difficult to overturn the decisions of the Appeals Board in court. The administrative penalty provisions are generally considered to provide the EPA with more leeway to issue and obtain administrative penalties than in the past, when the case went directly to a court of law.

The EPA can increase the $200,000 maximum, in consultation with the Department of Justice (DOJ), if such an increase appears appropriate. In addition, although such administrative penalties are limited by law to be filed within one year of the occurrence of the violation, the one-year limitation can be extended by the DOJ when appropriate.

ADMINISTRATIVE ORDERS

Administrative orders are orders issued to facilities requiring them to comply with certain provisions of the CAA. In the past, such orders could only be issued if the violation could be corrected within 30 days. Now, however, the orders can be issued for a period as long as one year. This allows the EPA more freedom to require specific actions backed up by administrative, civil, and criminal penalty provisions if the order is not complied with. These orders can be issued to require such things as auditing, additional records on control equipment, compliance certifications, and other administrative requirements that may or may not be related to specific permit or emission limit violations.

FIELD CITATIONS

This is an entirely new program under the 1990 CAA and allows EPA field inspectors to issue, on the spot, field citations up to $5000 per day per violation for alleged violations observed during the inspection. A field citation has been called a "toxic ticket" and is similar to a traffic violation citation. Once issued, the facility must pay the assessed penalty or appeal to the EPA through an informal hearing process. This informal hearing process is not a full-fledged evidentiary hearing, but rather a limited administrative hearing. If unsuccessful in the informal hearing process, the alleged violator can appeal to the federal district court, but, again, the appeal is limited to a review of the body of evidence considered by the EPA to ensure that the Agency did not err in issuing the citation.

CONTRACTOR LISTINGS

The 1990 CAA has expanded the EPA's contractor listing program. In the past, if a facility was in violation of the CAA, it could be listed by the EPA as a violator in the Federal Register and the U.S. government would be prohibited from doing business with that facility. Under the 1990 CAA, listing is mandatory for all criminal violations and the EPA can expand the list to include not only the violating facility, but also any other facility owned by the violator.

BOUNTY HUNTER PROVISIONS

Under the 1990 CAA, the EPA is authorized to award up to $10,000 to any person who furnishes information or services that lead to a criminal conviction or a judicial or administrative penalty for any violation of the Act. (Officers and employees of state and federal agencies cannot collect the award if the information is furnished in the performance of an official duty.)

CITIZEN SUITS

In the past, if citizens were not satisfied with Clean Air Act enforcement, the only type of suit they could file would be a suit to force the local permitting authority or the EPA to comply with the State Implementation Plan requirements or nondiscretionary duties required by the Act. In other words, the suits were limited to being against the agencies, not the facility.

Under the 1990 CAA, however, citizens can sue the individual facility to require compliance and collect penalties for noncompliance. The funds collected by such suits will be placed into a penalty fund that the EPA is to use for air compliance and enforcement activities. In addition, the courts can award fees to citizens to cover the cost of bringing the litigation. Citizen suits can be filed for past violations as well as current, ongoing violations of the Act.

CONFIDENTIAL INFORMATION

Faced with the above array of potential fines and enforcement possibilities, many managers are undertaking internal audits and compliance reviews. There is some concern, however, that if such a review is undertaken, information will be found that could "automatically" lead to a penalty. It is this author's opinion that self-discovery and reporting, even if reporting "bad news," is always better than the agency or public discovering and reporting potential violations. It is also this author's opinion that, when reporting the results of a review, it must be assumed that everything written will probably someday be made available to an enforcement agency. Therefore, when reporting the results of compliance reviews, the reporting must be as factual as possible. In addition, although it may be necessary to report the facts of a potential violation, it is not necessary to report or make available opinions related to the legal impact of such facts. Such legal interpretations and advice may be protected under the legal doctrines of attorney–client privilege and attorney work product. If protection of information is an issue, legal counsel should be sought before the potentially sensitive information is obtained in order to ensure that the information is properly handled.

SUMMARY

The new enforcement provisions of the 1990 CAA are varied, serious, and numerous. In addition, due to the new Title V operating permit program recordkeeping and annual certification requirements, the information and data needs necessary to file enforcement actions are being made readily available to the agencies and interested public. Accordingly, it is critical that facilities develop and implement a sound program of compliance review and tracking.

CHAPTER

14

TRANSPORTATION AND GENERAL CONFORMITY

INTRODUCTION

New Source Review (NSR, Chapter 8) and nonattainment area regulations (Chapter 9) generally apply mostly to stationary sources of emissions. The NSR regulations for attainment areas require that proposed new sources be evaluated for air quality impacts compared to significant deterioration increments and that the Best Available Control Technology (BACT) be installed on the source. The NSR regulations do not, by definition, apply to existing sources. In addition, except for a few special cases, NSR regulations do not consider nonstationary sources of emissions (such as vehicular traffic).

In nonattainment areas, new stationary sources must obtain offsets and install Lowest Achievable Emission Rate (LAER) technology, which is more stringent than BACT. Existing sources must install Reasonably Available Control Technology (RACT) as required and implemented by the State Implementation Plan (SIP). The SIP is prepared and issued by the local permitting authority (usually the state or a major political subdivision within the state, e.g., a county) and approved by the EPA. (For convenience, the term "state" will be used to refer to the state or local permitting authority that has responsibility for the air quality regulations in the local area.) The SIP details the set of regulations being used by the state to bring the area into attainment. The SIP includes RACT requirements, trip reduction plans, and other vehicle management programs to reduce the amount of vehicular traffic in the nonattainment area, requirements for emission offsets prior to construction of new sources, LAER requirements, and other programs that reduce the level of air pollution in the area.

The SIP is based upon an emission inventory developed by the state. The emission inventory identifies categories of sources, including mobile sources, and apportions the relative contribution of each source category to the overall nonattainment problem. Therefore, mobile sources are included in the emission inventory. However, other than programs that affect the amount of vehicle miles traveled (VMT), the SIP does not generally explicitly reduce vehicular emissions. Emissions from individual vehicles are regulated on a nationwide basis through tailpipe emission standards established pursuant to Title II of the Clean Air Act. In addition, in the past, emissions from the additional vehicles that will be added to an area as a result of the new source being built have not been generally considered in the stationary source permitting effort.

However, the transportation and general conformity regulations, promulgated November 24, 1993 (40 CFR Part 51 and Part 93) and following, changed that scenario and now require that the total emissions from additional vehicles related to a new transportation project (i.e., "transportation conformity") or a new major stationary source (i.e., "general conformity") be considered. The transportation and general conformity regulations require that, prior to any transportation-related project being funded (in the transportation conformity case) or prior to any new federal decision allowing a new stationary source to be built (the general conformity case), there must be a demonstration that the new project conforms to the SIP. In general, the term "conforms to the SIP" means that the new project will not make the air quality worse nor interfere with an area reaching attainment of the National Ambient Air Quality Standards (NAAQS). Currently, the 40 CFR Part 51 and 93 conformity regulations apply only in nonattainment and maintenance areas; however, the EPA plans to issue rules that will apply in attainment areas as well.

DEFINITION OF CONFORMITY

The legislative requirement for conformity demonstrations is stated in Section 176(c) of the CAA, which states the following:

(c)(1) No department, agency or instrumentality of the Federal Government shall engage in, support in any way or provide financial assistance for, license or permit, or approve, any activity which does not conform to an implementation plan [i.e., SIP]. . . .

The act further requires that assurance of conformity is an affirmative responsibility of the head of the federal department and requires the following elements in a conformity demonstration:

- The project (i.e., either a transpiration related project or a stationary source project) conforms to the SIP's purpose of eliminating or reducing the severity and number of violations and achieving expeditious attainment of the NAAQS.
- The project will not cause or contribute to any new violation of a NAAQS.
- The project will not increase the frequency or severity of any existing violation of a NAAQS.
- The project will not delay timely attainment of the NAAQS or any other required interim emission reductions or other milestones in any nonattainment area.

Note that there is a two-pronged test for conformity. One prong is that the conformity demonstration must show that the project is in conformance with the SIP. The second prong, which must not be ignored, is that the conformity demonstration must show that the project in and of itself will not cause NAAQS violations nor delay attainment. Conformity is a requirement on any federal agency, whether it normally controls air pollution or not. For example, if the Fish and Wildlife Service is funding a new fish hatchery, and the fish hatchery is located in a nonattainment area, the Fish and Wildlife Service would have to demonstrate that the project conforms to the SIP and does not cause any new violations of the NAAQS.

GENERAL CONFORMITY

From an industrial facility standpoint, general conformity is the most relevant type of conformity requirement. Unless exempted by regulation, the general conformity regulations apply to all projects that are not transportation related and require action on the part of some federal agency, including the EPA. The requirement that "any Federal agency action" must conform is very broad. For example, since the EPA approves all Environmental Impact Statements (EISs) prepared under the National Environmental Protection

Act (NEPA), if a project is required to undergo NEPA, it must also be evaluated for conformity.

Natural gas pipeline projects exemplify how broad the conformity requirement is. If a natural gas pipeline is to be constructed over public land, it will need an EIS and will need to obtain a right of way from the Bureau of Land Management (BLM) for building the pipeline. In addition, the Federal Energy Regulatory Commission (FERC) will have to approve the project. If any portion of the project enters into a nonattainment area, conformity would have to be demonstrated before the BLM could issue the right of way, before the EPA could approve the EIS, or before the FERC could approve the project. One can see how complicated conformity can get since, even in this simple example, three different federal agencies would have to issue a conformity determination. In fact, from a practical matter, interagency coordination is one of the biggest difficulties in obtaining an approved conformity demonstration.

Applicability

Other than transportation-related projects, general conformity applies, by regulation, to any federal action (not specifically exempted) that supports in any way (e.g., issuing a land use permit) a project where the total of direct and indirect emissions of the project in a nonattainment or maintenance area exceeds the thresholds shown in Table 14-1.

It is important to note in Table 14-1 that the thresholds are for the total of *direct* and *indirect* emissions. Contrary to the NSR program, where only the direct emissions of the source (and potential secondary emissions under the PSD program) are considered, the conformity regulations require evaluation of all emissions, including those from vehicles related to the project (e.g., employees traveling to the site). Therefore, it is possible for a source to be nonmajor under the NSR regulations (e.g., a well-controlled semiconductor manufacturing facility), but be major for purposes of conformity demonstrations (since the facility may have thousands of employees).

Direct emissions are emissions of a criteria pollutant or its precursors (see Chapter 4) "that are caused by or initiated by the Federal action and occur at the same time and place as the action" (40 CFR 51.852). Indirect emissions are emissions of a criteria pollutant or its precursors that are caused by the federal action but may occur later in time and/or may be farther removed in distance from the action. However, in order for indirect emissions to count, they must be (1) reasonably foreseeable and (2) practicably controllable due to a continuing responsibility of the federal agency.

TABLE 14-1 Nonattainment and Maintenance Area Conformity Applicability Thresholds (40 CFR 51.853)

Pollutant and Area Classification	Nonattainment Area (NAA) Thresholds (tons per year)	Maintenance Area Thresholds (tons per year)
VOCs or NO_x in serious ozone NAAs	50	NA[a]
VOCs or NO_x in severe ozone NAAs	25	NA
VOCs or NO_x in extreme ozone NAAs	10	NA
VOCs or NO_x in other ozone NAA outside an ozone transport region	100	100
VOCs in marginal and moderate NAAs inside an ozone transport region	50	50
NO_x in marginal and moderate NAAs inside an ozone transport region	100	100
CO in all NAAs	100	100
SO_2 in all NAAs	100	100
NO_2 in all NAAs	100	100
PM_{10} in moderate NAAs	100	100
PM_{10} in serious NAAs	70	100
Lead in all NAAs	25	25
Regionally significant (see text) projects—any criteria pollutant	More than 10% of total emissions of that pollutant in the specific NAA	More than 10% of total emissions of that pollutant in the specific maintenance area

[a]NA, not applicable.

In the semiconductor manufacturing plant example, direct emissions include VOCs and other emissions from the manufacturing plant stacks, while indirect emissions include emissions from vehicles of the employees, delivery trucks, and so on, traveling to and from the plant and fugitive emissions emitted at the plant (e.g., emissions from forklifts operating outside of the plant).

Exemptions from General Conformity

General conformity applies to *any* federal action. This is, obviously, a very broad term. Accordingly, the regulations provide an extensive list of exceptions. The exceptions, in general, include the following (consult 40 CFR 51.853 for a complete list of exemptions):

- Actions that do not increase emissions or increase emissions by only a *de minimis* amount, especially with respect to existing activities, renewing, and continuing actions (such as permit renewals)
- Actions of the Federal Reserve Bank
- Actions that implement a foreign affairs function of the United States
- Transfer of land or other real property
- Initial outer continental shelf leases
- Electric power marketing and acquisition, sale, and transmission of electric energy
- Actions to carry out a conforming program, such as prescribed burning
- Actions in response to emergency disasters
- Research and other investigations
- Alteration and additions to existing structures to comply with other environmental regulations (such as installation of scrubbers)
- CERCLA (Superfund-related emissions) as long as NSR is complied with
- Those portions of a project that are already regulated by NSR provisions

Demonstrating Conformity

There are two types of conformity demonstrations:

- Presumptive conformity
- Project-specific conformity

Presumptive conformity applies to types of projects that a specific federal agency has evaluated and determined that the projects meet the conformity tests of (1) not contributing to new violations, (2) not causing more or more severe violations, (3) not interfering with SIP progress to attainment or maintenance, or (4) not delay attainment, or that the projects always have emissions less than the applicability thresholds of Table 14-1. Before a presumptive conformity list can be finalized, the federal agency must conduct the analysis, publish its intentions in the Federal Register, and respond to comments.

Project specific conformity demonstrations must be conducted if the project:

- Has total emissions greater than the thresholds
- Is not exempted
- Is not on the presumptive conformity list (if an agency has, in fact, published such a list)
- Is not "regionally significant," regardless of whether or not the project emissions are greater than the thresholds of Table 14-1

Regionally significant projects are those projects whose total of direct and indirect emissions is greater than 10 percent of the total emissions of that pollutant in the nonattainment area or maintenance area (usually a state county or group of counties).

Elements of a Project-Specific Conformity Demonstration

In order for a conformity demonstration to be successful, the federal agency must demonstrate that the project is consistent with all other air quality requirements, such as NSR permitting, and that it meets *one or more* of the following options:

1. Demonstrate that the total project emissions are less than the applicability thresholds.
2. Demonstrate that the specific project emissions (total of direct and indirect) were *specifically* included and accounted for in the NAA or maintenance area SIP.
3. For VOC or NO_x emissions, completely offset total project emissions through other SIP-approved or similarly enforceable emission reductions in the area. (Note that this requirement is more stringent than just the stationary source requirements, because in conformity, both direct and indirect emissions must be offset, where in stationary source requirements, only direct emissions are offset.)
4. For pollutants other than NO_x and VOC, demonstrate through ambient air quality modeling that the total project emissions will not cause or contribute to any new exceedance of the NAAQS in the area (termed a *hot spot analysis*), nor increase the frequency or severity of existing exceedances.
5. For NO_x and VOC, demonstrate that the total project emissions will not exceed SIP emissions budgets, or if it would exceed the emissions budget, the following must occur:

a. A SIP revision to include the project and reduce emissions from other sources will be filed and approved prior to the project being initiated.
b. The SIP revision will specifically identify emission reductions.
c. All existing SIP requirements in the area are being implemented.
d. The federal agency has implemented all reasonable mitigation measures.

In addition to the technical analysis, there is extensive public participation and review of the conformity demonstration required as well as review and concurrence of all affected agencies. This aspect of the conformity demonstration is time-consuming and potentially fraught with possibilities for litigation and delay of the project.

Keys to Successful General Conformity Demonstrations

From a practical, industry perspective, the keys to success in conformity demonstrations are the following:

- Ensure that both direct and indirect emissions are counted in evaluating whether or not conformity applies.
- Track and participate in the local permitting authority's nonattainment and maintenance area air quality management plan (AQMP) and SIP submittals to ensure that the emissions inventory includes anticipated emissions that could occur should the facility expand in the future.
- Ensure that the emissions so identified are explicitly presented in the AQMP and emissions inventories.

TRANSPORTATION CONFORMITY

The concept of transportation conformity is the same as general conformity, but it applies to federal funding of transportation related projects. The elements of the conformity demonstration are the same as general conformity. From an industry perspective, transportation conformity is not likely to impact facility compliance issues.

SUMMARY

The current comprehensive conformity regulations are relatively new and, to date, have not become a severe burden for industry. However, the amount of public participation required by the conformity regulations, the difficulty of demonstrating that a project's emissions are included in the local emissions inventory, and the fact that conformity requirements are overlain on top of all other environmental requirements indicate that conformity could become a major hurdle preventing industrial expansion in the future. This is especially the case once the EPA publishes its attainment area conformity regulations. Therefore, even though this portion of the CAA has not yet been a major impediment to most industrial facilities, it is an issue that must be tracked and incorporated into air quality compliance management.

CHAPTER

15

FUTURE IMPLEMENTATION OF THE 1990 CAA

The 1990 Clean Air Act Amendments required the EPA to develop and promulgate more than 30 major new regulatory programs in the first three years after enactment (i.e., prior to November 15, 1993). The Agency worked diligently to reach that milestone, but it is behind schedule for most programs. The programs that are the most delayed are those requiring new technological determinations related to emission limits (e.g., the MACT standards for hazardous air pollutant source categories).

The various programs and their timing have been discussed in previous chapters of this *Guide*. However, for reference and as a tool to identify new programs, Table 15-1 shows the statutory schedule for CAA rule development.

TABLE 15-1 Schedule of CAA Implementation

Strategy Target Date	Title	Activity	Completed	Statutory Deadline
Dec 1990	Title I—Nonattainment	Issue "Getting Started" Letter to Governors	Y	
		States submit request/justification for 5 percent classification adjustments	Y	Dec 1990
Jan 1991	Title I—Nonattainment	Publish two-year implementation strategy	Y	
		Publish notice of initial PM_{10}, SO_2, and lead nonattainment areas	Y	
	Title VI—CFCs	Prepare listing of ozone depleting substances	Y	Jan 1991
Feb 1991	Title I—Nonattainment	Act on 5 percent classification adjustment requests	Y	Feb 1991
Mar 1991	Title I—Nonattainment	States designate nonattainment areas	Y	Mar 1991
	Title III—Air Toxics	Propose early reductions rulemaking	Y	
	Title VI—CFCs	Direct final 1991 production limits	Y	
Apr 1991	Title I—Nonattainment	States submit PM_{10} areas unable to reach attainment by 1994	Y	
		States respond to list of PM_{10}, SO_2, and lead nonattainment areas	Y	
	Title V—Permits	Propose state permit regulations	Y	
May 1991	Title I—Nonattainment	States complete RACT corrections	Partial	May 1991
		Notify states of intent to modify suggested boundaries	Y	May 1991
	Title II—Mobile Sources	Finalize gasoline Reid Vapor Pressure regulations	Y	May 1991
		Finalize Tier I car and truck standards	Y	May 1991

TABLE 15-1 *(Continued)*

Strategy Target Date	Title	Activity	Completed	Statutory Deadline
		Propose reformulated gasoline requirements	Y	
		Propose clean fuels fleet and California pilot credit programs	Y	
		Propose urban bus regulations	Y	
		Propose emission control diagnostics rule	Y	
	Title IV—Acid Rain	Propose regulations for auctions and sales	Y	
June 1991	Title I—Nonattainment	Propose PM_{10} area reclassifications	Y	June 1991
	Title III—Air Toxics	Publish draft list of source categories	Y	
		Propose list of high-risk pollutants, 90/95 percent early reductions guidance	Y	
July 1991	Title I—Nonattainment	Finalize ozone, CO, PM_{10}, and lead nonattainment boundaries	Y	July 1991
	Title II—Mobile Sources	Propose lead substitute gasoline additives	Y	
	Title VII—Enforcement	Propose administrative penalties rules of practice	Y	
Sept 1991	Title VI—CFCs	Propose CFC phaseout regulations	Y	
		Propose mobile air conditioners recycling regulations	Y	
Oct 1991	Title I—Nonattainment	Publish VOC and CO emission inventory guidance	Y	
	Title II—Mobile Sources	Publish study on nonroad engines	Y	Nov 1991
Nov 1991	Title I—Nonattainment	Publish guidance on control cost effectiveness	Y	Nov 1991
		States submit PM_{10} SIPs	Partial	Nov 1991
		Publish air quality and emissions trends report	Y	

TABLE 15-1 Schedule of CAA Implementation (*Continued*)

Strategy Target Date	Title	Activity	Completed	Statutory Deadline
		Establish Grand Canyon Visibility Transport Commission	Y	Nov 1991
	Title III—Air Toxics	Propose MACT for dry cleaners (per court order)	Y	
Dec 1991	Title I—Nonattainment	Propose outer continental shelf rule	Y	
	Title IV—Acid Rain	Finalize regulations for auctions and sales	Y	Nov 1991
		Propose allowance trading system	Y	
		Propose acid rain permit program	Y	
		Propose continuous emission monitor requirements	Y	
		Propose excess emissions rules	Y	
		Propose conservation and renewable incentives	Y	
Jan 1992	Title V—Permits	Publish guidance on state programs to assist small businesses	Y	
	Title VI—CFCs	Propose ban on nonessential products	Y	
Feb 1992	Title II—Mobile Sources	Finalize urban bus regulations (1991 and 1992)	Y	
	Title VII—Enforcement	Finalize administrative penalties rules of practice	Y	
Mar 1992	Title I—Nonattainment	Publish Title I General Preamble	Y	
	Title II—Mobile Sources	Finalize onboard controls regulatory decision	Y	Nov 1991
Apr 1992	Title I—Nonattainment	Propose rules for ozone, NO_x, and VOC monitoring	Y	
	Title II—Mobile Sources	Propose rule on reformulated gasoline SNPRM (supplemental notice of proposed rulemaking)	Y	Nov 1991

■ **TABLE 15-1** *(Continued)*

Strategy Target Date	Title	Activity	Completed	Statutory Deadline
May 1992	Title I—Nonattainment	Publish guidance on TCMs (Transportation Control Measures)	Y	Nov 1991
		Convene NE Ozone Transport Commission	Y	May 1991
	Title II—Mobile Sources	Finalize cold temperature CO standards	Y	Nov 1991
	Title VI—CFCs	Propose CFC labeling regulations	Y	
June 1992	Title V—Permits	Finalize state permit regulations	Y	Nov 1991
	Title VI—CFCs	Finalize mobile air conditioning recycling regulations	Y	Nov 1991
July 1992	Title I—Nonattainment	Propose discretionary sanction rules	Y	
		States submit SIPs of initial SO_2 nonattainment areas	Partial	May 1992
	Title II—Mobile Sources	Propose enhanced I/M regulation	Y	
	Title III—Air Toxics	Publish final list of source categories	Y	Nov 1991
	Title IV—Acid Rain	Propose list of Phase II utility allowances	Y	Dec 1991
	Title VI—CFCs	Finalize CFC phaseout regulations	Y	Sept 1991
		Revise two-year implementations strategy	Y	
Aug 1992	Title I—Nonattainment	Publish Title I NO_x guidance	Y	
		Finalize outer continental shelf rules	Y	Nov 1991
		Propose additional PM_{10} nonattainment areas	Y	
	Title II—Mobile Sources	Finalize oxygenated fuel credit guideline	Y	Aug 1991
		Finalize California pilot credit program	Y	Nov 1991

■ TABLE 15-1 Schedule of CAA Implementation (*Continued*)

Strategy Target Date	Title	Activity	Completed	Statutory Deadline
	Title III—Air Toxics	Propose HAP list petition procedures	Y	
		Propose rules for risk management plans and prevention programs	Y	
		Propose list of substances for accidental releases prevention program	Y	
	Title IV—Acid Rain	Propose NO_x requirements for utility boilers	Y	
	Title VI—CFCs	Propose emission reduction program	Y	
	Title VII—Enforcement	Propose rules for citizen suits	Y	
		Propose monetary awards rules	Y	
Sept 1992	Title I—Nonattainment	Issue transportation planning guidance (EPA/DOT)	Y	
		Finalize guidance on transportation air quality planning	Y	Aug 1991
		Finalize PM_{10} area reclassifications	Y	Dec 1991
		Propose economic incentive rules	Y	
		Finalize PM_{10} BACM technical guidance	Y	May 1991
	Title II—Mobil Sources	Finalize clean fuels fleet (TCMs)	Y	Nov 1991
		Propose revisions to motor vehicle certification procedures (I/M short test)	Y	
	Title III—Air Toxics	Propose regulatory schedule for all source categories	Y	
		Finalize early reductions rulemaking	Y	

■ **TABLE 15-1** *(Continued)*

Strategy Target Date	Title	Activity	Completed	Statutory Deadline
		Propose standards for large MWCs	Y	
		Propose standards for small MWCs	Y	
		Propose general provisions for MACT standards (40 CFR Part 63)	Y	
		Propose MACT for hazardous organic chemical manufacturing	Y	
	Title VI—CFCs	Propose accelerated CFC phaseout	Y	
	Title VII—Miscellaneous	Publish section 811 report (International Competitiveness)	Y	May 1992
Oct 1992	Title I—Nonattainment	Publish 1991 air quality data and emission trends report	Y	
		Propose transportation conformity criteria	Y	
	Title II—Mobile Sources	Propose clean fuel fleet vehicle standards, conversions, and general provisions	Y	Nov 1992
	Title IV—Acid Rain	Finalize excess emission requirements	Y	May 1992
		Finalize allowance trading system	Y	May 1992
		Finalize continuous emission monitoring requirements	Y	May 1992
		Finalize conservation and renewable resource incentives	Y	May 1992
		Finalize acid rain permit program	Y	May 1992
Nov 1992	Title I—Nonattainment	States submit RACT catch-up rules, NSR rules, CO attainment demonstrations and contingency measures	Y	Nov 1992

■ **TABLE 15-1** Schedule of CAA Implementation (*Continued*)

Strategy Target Date	Title	Activity	Completed	Statutory Deadline
		States submit small business assistance programs	Y	Nov 1992
		States submit base ozone, CO emission inventories	Y	Nov 1992
		States receive emission source statements	Y	Nov 1992
		PM_{10} PSD increments (11/92 court agreement)	Y	
	Title II—Mobile Sources	Propose reformulated gasoline complex model	Y	
		Publish draft mobile source air toxics study	Y	
		Finalize urban bus 94+ (pm standard)	Y	Jan 1992
		Finalize rules for lead substitution of gasoline additives	Y	Nov 1991
		Finalize enhanced I/M regulation	Y	Nov 1991
	Title III—Air Toxics	Complete study of hydrogen sulfide emission from extracting natural gas and oil	Y	Nov 1992
		Complete study of the hazards of hydrofluoric acid	Y	Nov 1992
		Finalize MACT for dry cleaners (per court order)	Y	Nov 1992
		Propose MACT for coke ovens	Y	
	Title VI—CFCs	Propose safe alternatives	Y	
	Title VII—Miscellaneous	Issue report on the impact of the 1990 CAA Amendments on visibility in Class I areas	Y	Nov 1992
Dec 1992	Title I—Nonattainment	Finalize additional PM_{10} nonattainment areas	Y	
		Publish alternative control technology documents for VOCs	Y	Nov 1993

■ **TABLE 15-1** (*Continued*)

Strategy Target Date	Title	Activity	Completed	Statutory Deadline
	Title II—Mobile Sources	Finalize emission control diagnostic rule	Y	May 1992
		Publish procedures of enforcement for urban bus standards	Y	
	Title VI—CFCs	Finalize ban on nonessential products	Y	Nov 1991
		Finalize accelerated CFC phaseout	Y	
Jan 1993	Title I—Nonattainment	Approve/disapprove first PM_{10} SIPs		Jan 1993
		Propose tank vessel rule	Y	
	Title II—Mobile Sources	Submit transportation systems report to Congress	Y	Jan 1993
		Finalize urban bus retrofit	Y	Nov 1991
	Title III—Air Toxics	Propose MACT for chromium electroplating	Y	
	Title VI—CFCs	Propose procurement regulations	Y	
	Title VII—Enforcement	Propose rules for contractor listing	Y	
		Propose rules for field citation program	Y	
Feb 1993	Title II—Mobile Sources	Finalize reformulated gasoline regulations	Y	Nov 1991
	Title III—Air Toxics	Propose MACT for commercial sterilizers	Y	
		Finalize list of substances for accidental releases prevention program	Y	Nov 1992
	Title IV—Acid Rain	Sources submit Phase I permit applications and proposed compliance plans	Y	Feb 1993
		Propose OPT-IN regulations—combustion sources	Y	
		Finalize list of Phase II utility allowances	Y	Dec 1992

■ **TABLE 15-1** Schedule of CAA Implementation (*Continued*)

Strategy Target Date	Title	Activity	Completed	Statutory Deadline
	Title VI—CFCs	Finalize CFC labeling regulations	Y	May 1992
	Title VII—Enforcement	Propose rules for enhanced monitoring and compliance certification	Y	
Mar 1993	Title I—Nonattainment	Finalize economic incentives rules	Y	Nov 1992
		Publish SOCMI distillation processes CTG	Y	Nov 1993
		Propose general conformity criteria	Y	
		Publish SOCMI reactor processes CTG	Y	Nov 1993
	Title II—Mobile Sources	Finalize reformulated gasoline complex model	Y	
		Publish mobile-source related air toxics study	Y	May 1992
		Determine significance of nonroad engine emissions and conduct rulemaking	Y	Nov 1992
		Finalize revisions to motor vehicle certification procedures (I/M short test)	Y	Nov 1991
	Title III—Air Toxics	Propose guidance for modification provisions	Y	
		Propose guidance for state air toxics program	Y	
	Title IV—Acid Rain	Initiate auctions and sales of allowances	Y	Mar 1993
Apr 1993	Title I—Nonattainment	States submit lead SIPs (18 months from designation)	Y	July 1993
		Promulgate sanctions for states failing to submit RACT fix-up rules (18 months from findings)		Apr 1993

TABLE 15-1 (*Continued*)

Strategy Target Date	Title	Activity	Completed	Statutory Deadline
		Finalize rules for enhanced monitoring of ozone, NO_x, VOCs		May 1992
	Title II—Mobile Sources	Finalize procedures of enforcement for urban bus standards	Y	May 1992
	Title III—Air Toxics	Propose State I gasoline marketing MACT	Y	
May 1993	Title I—Nonattainment	Finalize discretionary sanctions rules	Y	
		Propose revisions to NSR program	Y	
	Title II—Mobile Sources	Complete testing protocols for fuels and additives	Y	
	Title VI—CFCs	Finalize emission reduction program	Y	Jan 1992
June 1993	Title VII—Enforcement	Finalize rules for citizen suits	Y	
		Finalize rules for monetary awards	Y	
July 1993	Title I—Nonattainment	Publish 185B NO_x/VOC study	Y	Feb 1992
	Title II—Mobile Sources	Finalize clean fuel vehicle standards, conversions, and general provisions	Y	Nov 1992
	Title III—Air Toxics	Finalize HAP list petition procedures	Y	
		Finalize general provisions for MACT standards		
Aug 1993	Title I—Nonattainment	Finalize transportation conformity criteria		Nov 1991
	Title II—Mobile Sources	Finalize vehicle evaporative emissions regulations	Y	Aug 1993
	Title III—Air Toxics	Finalize MACT for coke ovens	Y	Dec 1992

■ TABLE 15-1 Schedule of CAA Implementation (*Continued*)

Strategy Target Date	Title	Activity	Completed	Statutory Deadline
	Title IV—Acid Rain	EPA action on Phase I permits	Y	Aug 1993
		Finalize NO_x requirements for utility boilers		May 1992
Sept 1993	Title III—Air Toxics	Finalize regulatory schedule for all source categories	Y	Nov 1992
		Propose medical waste incinerator rule	Y	
	Title VI—CFCs	Finalize safe alternatives program	Y	Nov 1992
Oct 1993	Title I—Nonattainment	Promulgate FIPs for states failing to submit RACT fix-up rules (2 years from findings)		
		Publish air quality and emission trends report	Y	
	Title III—Air Toxics	Finalize MACT for hazardous organic chemical manufacturing	Y	Nov 1992
		Propose MACT for pulp and paper	Y	
		Finalize standards for large MWCs	Y	Nov 1991
		Finalize standards for small MWCs	Y	Nov 1992
		Propose MACT for polymers and resins II	Y	
	Title IV—Acid Rain	Begin issuing allowances to small diesel refineries	Y	Oct 1993
		Propose OPT-IN regulations—process sources	Y	
	Title VI—CFCs	Finalize procurement regulations	Y	May 1992
Nov 1993	Title I—Nonattainment	Submit consumer/commercial products study report to Congress	Y	Nov 1993

■ **TABLE 15-1** (*Continued*)

Strategy Target Date	Title	Activity	Completed	Statutory Deadline
		Publish alternative control technology documents for NO_x sources	Y	Nov 1993
		Publish results of ozone design value study	Y	Nov 1993
		States submit selected SIP provisions for moderate and above ozone areas	Y	Nov 1993
		Publish auto body refinishing CTG	Y	Nov 1993
		Publish batch processes CTG	Y	Nov 1993
		Publish petroleum/industrial wastewater CTG	Y	Nov 1993
		Publish plastic parts coating CTG	Y	Nov 1993
		Publish storage tanks CTG	Y	Nov 1993
		Publish web offset lithography CTG	Y	Nov 1993
		Publish SOCMI batch processes CTG	Y	Nov 1993
		Publish surface coal mining study		Nov 1993
		Finalize general conformity criteria		Nov 1991
	Title III—Air Toxics	Submit Great Lakes study report to Congress	Y	Nov 1993
		Publish urban area strategy—research report	Y	Nov 1993
		Submit NAS study to Congress	Y	May 1993
		Finalize rules for risk management plans and prevention programs		Nov 1993
		Publish guidance for state air toxics programs		Nov 1991
	Title IV—Acid Rain	States submit acid rain permit program proposals	Y	Nov 1993

■ TABLE 15-1 Schedule of CAA Implementation (*Continued*)

Strategy Target Date	Title	Activity	Completed	Statutory Deadline
		Promulgate SO_2 new source performance standards (NSPS) for new fossil fuel utility units		Nov 1993
	Title V—Permits	States submit operating permit programs		Nov 1993
	Title VII—Miscellaneous	Publish interim source/receptor report on visibility		
Dec 1993	Title VII—Enforcement	Finalize rules for contractor listing	Y	
		Finalize guidance/rules for field citation program		

Source: After USEPA (1992a).

Further detail regarding the status of the various regulatory programs can be found by consulting EPA's regulatory agenda, which is published in the Federal Register on a semiannual basis (Chapter 3). In addition, if one needs to obtain even further detail, the EPA maintains a computer bulletin board that contains a very large amount of current information related to promulgated and in-work projects and rules. This bulletin board is maintained by the EPA Office of Air Quality Planning and Standards (OAQPS) and can be accessed by dialing 919-541-5742 (modem) or 919-541-5384 (voice). For computer access, the bulletin board is set for 8-bit, no parity, full duplex at data transfer speeds over 14,400 bits per second (baud) and is available free of charge.

Finally, complete details for each program of interest are found in the Federal Register and/or by contacting the assigned individuals at the EPA. In addition, as new regulatory programs are promulgated and/or new interpretions of the regulations implemented, this *Guide* will be updated to incorporate the most current information.

CHAPTER 16

SUMMARY

The 1990 Clean Air Act Amendments (CAA) have placed on industry a massive and very complex new set of regulatory requirements. In addition, the Act has significantly increased the penalties associated with violating any provisions of these new regulatory requirements. In order to comply, the environmental manager must not only know what the specific regulatory requirements are, but also develop an understanding of the fundamental underlying concepts of the regulations. This *Guide* has attempted to provide useful information in both of these areas in an understandable and readable format.

In order to achieve compliance, the following seven-step process is recommended to affected facilities. The recommended process is a bottoms-up approach to compliance that starts with understanding and ends with upper management and corporate commitment. The process is as follows:

1. Using this *Guide* and other reference documents, understand what the Act is attempting to accomplish with respect to a certain regulatory program (e.g., stratospheric ozone protection, nonattainment, new source review, etc.) that could affect the facility.

2. Understand how the requirements could potentially affect the facility by comparing facility operations to the overall goals of the regulatory programs.
3. Research the specific details of the requirements, and determine precisely the applicability and specific regulatory details of the program.
4. Match the specific demands of the requirements to the specific operational needs of the facility, both current and future.
5. Repackage the regulatory and operational information into a format that is understandable to upper management, and impress upon them the seriousness of complying with the requirements.
6. Develop an operational and regulatory strategy to achieve compliance.
7. Implement the strategy with adequate follow-up and review.

If the above steps are carried out, the new regulatory requirements can be incorporated into the overall strategic and tactical business plan for the facility. In this manner, air quality compliance is more easily achieved and expenditures minimized.

When the regulatory strategic and tactical plan has been developed and implemented, it is necessary to conduct routine internal compliance reviews in order to help ensure that long-term and continuous compliance is achieved. These reviews (sometimes referred to as environmental audits) can be conducted with internal or outside assistance. The key is not necessarily who does the review, but that it is done and done consistently. The reason that this author uses the term "compliance reviews" instead of "audits" is that the certainty and precision normally associated with the term "audit" (as in an accounting audit) cannot be achieved in the air quality arena. There are simply too many different interpretations of too many different regulations in order to provide that level of certainty. However, regardless of the fact that certainty cannot be achieved, conducting such reviews is a critical management requirement and should be implemented on a routine and regular basis. The results of the review should be forwarded to the highest levels of management.

When conducting environmental compliance reviews, there are often a number of gray areas where it is uncertain how a specific regulation applies or does not apply to the topic in question. In that case, the facility can consult outside legal counsel, outside technical consultants, and/or the agencies themselves. If the questions are specific and properly phrased (see, for example, the Ten Steps for Effective Regulatory Process in Chapter 3), answers can be obtained from the agencies that will yield an adequate degree of confidence.

Congressman John Dingell, one of the sponsors of the 1990 Clean Air Act Amendments, called the Act the most complex, comprehensive, and far-reaching environmental law any Congress has ever considered. Senator Max Baucus, the leading spokesman for the Senate congressional conference committee, termed the Amendments the most sweeping environmental law that Congress will pass this century (CCH, 1990). It is easy to see how one can become overwhelmed and be filled with despair when attempting to decipher the regulatory requirements of the Act. It is hoped that this *Guide* has allowed the reader to shed some of the burden he or she is carrying as the road to air quality compliance is traveled for the remainder of this century and beyond.

REFERENCES

ABA. 1993. *Update: Implementing the 1990 Clean Air Act.* American Bar Association, February 11, 1993.

ACGIH. 1993. *Threshold Limit Values and Biological Indices for 1992–1993*, fifth printing. American Conference of Governmental Industrial Hygienists.

AICHE. 1992. *Guidelines for Hazard Evaluation Procedures*, second edition. American Institute of Chemical Engineers.

AWMA. 1991. *New Source Review Prevention of Significant Deterioration and Nonattainment Guidance Notebook.* Air and Waste Management Association.

Blattner, J. Wray. 1992. *The Clean Air Act Compliance Handbook.* Executive Enterprises Publications Co., Inc.

CAPCOA. 1989. *Air Toxics Assessment Manual.* California Air Pollution Control Officers Association.

CAPCOA. 1990. *Air Toxics Hot Spots Program Facility Prioritization Guidelines.* California Air Pollution Control Officers Association, July 1990.

CAPCOA. 1991. *Air Toxics Hot Spots Risk Assessment Guidelines.* California Air Pollution Control Officers Association, July 1991.

CAPCOA. 1992. *Air Toxics "Hot Spots" Program Risk Assessment Guidelines.* California Air Pollution Control Officers Association.

CCH. 1990. *Clean Air Act Law and Explanation.* Commerce Clearing House, Inc.

Elliott, E. Donald, Richard M. Schwartz, Alan V. Goldman, Alan B. Horowitz, and Joseph Laznow. 1992. "The Clean Air Act: New Enforcement and Liability Provisions," *Journal of the Air and Waste Management Association*, Volume 42, No. 11, November 1992.

Erbes, Russell E. 1993. "Principles for an Effective Regulatory Process," *Journal of Environmental Regulation*, Summer, 1993.

Hanna, Steven R., Gary A. Briggs, and Rayford P. Hosker, Jr. 1982. *Handbook on Atmospheric Diffusion.* U.S. Department of Energy Technical Information Center.

SCAQMD. 1988. *Best Available Control Technology Guidance Manual.* South Coast Air Quality Management District.

SCAQMD. 1993. *Rules and Regulations.* South Coast Air Quality Management District.

Slade, David H., ed. 1968. *Meteorology and Atomic Energy.* U.S. Atomic Energy Commission Office of Information Services.

STP. 1990. *The Complete Guide to Hazardous Materials Enforcement and Liability.* Specialty Technical Publishers, Inc.

Turner, D. Bruce. 1970. *Workbook of Atmospheric Dispersion Estimates.* U.S. Environmental Protection Agency, Office of Air Programs.

Turner, D. Bruce. 1994. *Workbook of Atmospheric Dispersion Estimates,* second edition. Lewis Publishers, CRC Press Inc.

USDA. 1992. *Guidelines for Evaluating Air Pollution Impacts on Class I Wilderness Areas in California.* USDA Forest Service General Technical Report PSW-GTP-136.

USEPA. 1990. *New Source Review Workshop Manual.* U.S. Environmental Protection Agency, Office of Air Quality Planning and Standards, Research Triangle Park.

USEPA. 1992a. *Implementation Strategy for Clean Air Act Amendments, 1992 Update,* U.S. Environmental Protection Agency. Published by the Bureau of National Affairs, Inc., Washington, DC, August 28, 1992.

USEPA. 1992b. Letter from William Rosenberg, assistant administrator, to William Riley, administrator, November 12, 1992. Letter was reprinted by the American Bar Association in *Update: Implementing the 1990 Clean Air Act,* February 11, 1993.

USEPA. 1993. Policy memorandum written by Lydia Wegman, available from USEPA Bulletin Board (TTN).

USOMB. 1987. *Standard Industrial Classification Manual.* U.S. Office of Management and Budget. Published by the National Technical Information Service.

APPENDIX A

LIST OF ACRONYMS

AEL	Acceptable exposure level
AEP	Actual exposed person
AHM	Acutely hazardous material
AQCR	Air quality control region
AQIA	Air quality impact analysis
AQRV	Air quality related values
ATC	Authority to construct
BACM	Best achievable control measures
BACT	Best available control technology
BTU	British thermal unit
CAA	Clean Air Act
CAAA	Clean Air Act Amendments
CFC	Chlorofluorocarbon
CFR	Code of federal regulations
CMSA	Consolidated metropolitan statistical area
CTG	Control technique guideline
EHS	Extremely hazardous substance
ERC	Emission reduction credit
FGD	Flue gas desulfurization
FIP	Federal implementation plan
FR	Federal Register
g/s	Grams per second
GACT	Generally available control technology
HAP	Hazardous air pollutant
HAZAL	Hazard Analysis
HAZOP	Hazard and operability study
HBFC	Hydrobromofluorocarbon
HCFC	Hydrochlorofluorocarbon
HI	Hazard index
HON	Hazardous organic national emission standard for hazardous air pollutants (i.e., hazardous organic NESHAP)

HRA	Health risk assessment
IPP	Independent power producer
LAER	Lowest achievable emission rate
lb/mmBTU	Pounds per million BTU
LOC	Level of concern
MACT	Maximum achievable control technology
MEF	Major emitting facility
MEIR	Maximum exposed individual at a residence
MEIW	Maximum exposed individual at a place of work
mg	milligrams (10^{-3} grams)
microns	One-millionth of a meter (10^{-6} meters)
MSA	Metropolitan statistical area
NA	Nonattainment
NAAQS	National ambient air quality standards
NESHAP	National emission standard for hazardous air pollutants
NIOSH	National Institute for Occupational Safety and Health
NMHC	Nonmethane hydrocarbons
NMOG	Nonmethane organic gases
NSPS	New source performance standards
NSR	New Source Review
NTIS	National Technical Information Service
OCA	Offsite consequence analysis
ODC	Ozone depleting compound
ODP	Ozone depleting potential
OMB	United States Office of Management and Budget
PCE	Perchloroethylene
PM	Particulate matter
PM_{10}	Particulate matter with mean aerodynamic diameter less than 10 microns
ppb	Parts per billion (10^{-9})
ppm	Parts per million (usually by volume—ppm(v), 10^{-6})

PSD	Prevention of significant deterioration
PTC	Permit to construct
PTO	Permit to operate
RA	Risk assessment
RACM	Reasonably available control measures
RACT	Reasonably achievable control technology
REG-NEG	Regulatory negotiation
RFP	Reasonable further progress
RMPP	Risk Management and Prevention Program
ROC	Reactive organic compounds
ROG	Reactive organic gases
SAAQS	State ambient air quality standards
SARA	Superfund Amendments and Reauthorization Act
SCR	Selective catalytic reduction
SIC	Standard Industrial Classification
SIP	State Implementation Plan
TCA	1,1,1-Trichloroethane
TCM	Transportation control measures
TIP	Tribal Implementation Plan
TLV	Threshold limit value
TNMHC	Total nonmethane hydrocarbons
TPQ	Threshold planning quantities
TRI	Toxic release inventory
TRS	Total reduced sulfur
TSP	Total suspended particulate
μg	Micrograms (10^{-6} grams)
$\mu g/m^3$	Micrograms per cubic meter
USEPA	United States Environmental Protection Agency
VMT	Vehicle miles traveled
VOC	Volatile organic compound

APPENDIX B

GLOSSARY OF TERMS

Acid Deposition ("Acid Rain"). A complex chemical and atmospheric phenomenon that occurs when emissions of sulfur and nitrogen compounds and other substances are transformed by chemical processes in the atmosphere, often far from the original sources, and then deposited on earth in either a wet or dry form. The wet forms, popularly called "acid rain," can fall as rain, snow, or fog. The dry forms are acidic gases or particulates.

Acute Exposure. Short-term exposure to a toxic substance. Acute exposures are generally characterized as lasting no longer than a day and often as one-hour exposures.

Afterburner. In incinerator technology, a burner located so that the combustion gases are made to pass through its flame in order to remove smoke and odors. It may be attached to or be separated from the incinerator proper.

Air Board. General term normally used to designate the local or state air agency responsible for permitting and enforcement of permits.

Air Contaminant. Any particulate matter, gas, or combination thereof, other than water vapor or natural air. (See "Air Pollutant.")

Air Pollutant. Any substance in air that could, if in high enough concentration, harm humans, other animals, vegetation, or material. Pollutants may include almost any natural or artificial composition of matter capable of being airborne. They may be in the form of solid particles, liquid droplets, gases, or in combinations of these forms. Generally, they fall into two main groups: (1) those emitted directly from identifiable sources; and (2) those produced in the air by interaction between two or more primary pollutants, or by reaction with normal atmospheric constituents, with or without photoactivation.

Air Quality Control Region. An area designated by the federal government in which communities share a common air pollution problem. Sometimes several states are involved.

Air Toxics. Any air pollutant for which a national ambient air quality standard (NAAQS) does not exist (i.e., excluding ozone, carbon monoxide, PM_{10}, sulfur dioxide, nitrogen dioxide, and lead) that may reasonably be anticipated to cause cancer, developmental effects, reproductive dysfunctions, neurological disorders, heritable gene mutations, or other serious or irreversible chronic or acute health effects in humans.

Algorithm. A specific mathematical calculation procedure. A model may contain several algorithms.

Ambient Air. Any unconfined portion of the atmosphere to which the public has access (i.e., outside the fenceline of a facility).

Attainment Area. An area considered to have air quality as good as or better than the national and local ambient air quality standards. An area may be an attainment area for one pollutant and a nonattainment area for others.

Authority to Construct (ATC). The permit issued to a user before construction begins on a new source of emissions.

Background. Ambient pollutant concentrations due to (1) natural sources, (2) nearby sources other than the one(s) currently under consideration, and (3) unidentified sources.

Baghouse Filter. Large fabric bag, usually made of glass fibers, used to eliminate intermediate and larger (greater than 20 microns in diameter) particles. This device operates in a way similar to the bag of an electric vacuum cleaner.

Banking. A system for recording qualified air emission reductions for later use in bubble, offset, or netting transactions. (See "Emissions Trading.")

Base Case. The starting point for BACT analysis under the traditional PSD process, the control strategy that, absent the BACT requirement, would be adopted for the new or modified source.

Best Achievable Control Measure (BACM). A term used in the 1990 CAA referring to the "best" measures (according to EPA guidance) for controlling small or dispersed sources of particulate matter, such as roadway dust, woodstoves, and open burning.

Best Available Control Technology (BACT). An emissions limitation based on the maximum degree of reduction for each pollutant that would be emitted from a proposed major source or major modification that the EPA, on a case-by-case basis, taking into account energy, environmental, and economic impacts and other costs, determines is achievable for such source. (The BACT determination is a key component of the PSD program.)

Bubble Policy (or Emissions Trading). An approach to emissions limitations whereby limitations are applied to an entire source, rather than on individual emission units, as though an imaginary "bubble" were placed over the plant with but one point of emissions.

Calm. For purposes of air quality modeling, calm is used to define the situation when the wind is indeterminate with regard to speed or direction, usually less than about 0.5 meters per second.

Catalytic Incinerator. A control device that oxidizes volatile organic compounds (VOCs) by using a catalyst to promote the combustion process. Catalytic incinerators require lower temperatures than conventional thermal incinerators, with resultant fuel and cost savings.

CFCs (Chlorofluorocarbons). A family of inert, nontoxic, and easily liquefied chemicals used in refrigeration, air conditioning, packaging, insulation, or as solvents or aerosol propellants. Because CFCs are not destroyed in the lower atmosphere, they drift into the upper atmosphere where the chlorine is released and reacts with ozone.

Chronic Exposure. Long-term continuous exposure. (See "Acute Exposure.") Chronic exposures are usually characterized as annual exposures.

Clean Coal Technology. Any technology not in widespread use as of the date of enactment of the Clean Air Act that will achieve significant reductions in pollutants associated with the burning of coal.

Clean Fuels. Blends and/or substitutes for gasoline fuels. These include compressed natural gas, methanol, ethanol, and others.

Coke Oven. An industrial process that converts coal into coke, which is one of the basic materials used in blast furnaces for the conversion of iron ore into iron.

Cold Temperature CO. A standard for automobile emissions of carbon monoxide (CO) to be met at a low temperature (i.e., 20°F). Conventional catalytic converters are less efficient upon start-up at low temperatures.

Complex Terrain. Terrain exceeding the height of the stack being modeled.

Compliance Schedule. A negotiated agreement between a pollution source and a government agency that specifies dates and procedures by which a source will reduce emissions and, thereby, comply with a regulation.

Conformity. When applied to projects in nonattainment areas, the project emissions are consistent with a nonattainment area's plan to reach attainment.

Control Techniques Guideline (CTG). Guidance documents issued by the EPA that define reasonably available control technology (RACT) to be applied to existing facilities that emit certain threshold quantities of air pollutants. The CTGs contain information both on the economic and technological feasibility of available techniques.

Conventional Pollutants. Six common criteria pollutants: SO_x, NO_x, PM_{10}, lead, CO, O_3. (See "Hazardous Air Pollutants.")

Criteria Document. An EPA report setting forth the most recent scientific knowledge of the identifiable effects and reactive qualities of a particular pollutant for which a National Ambient Air Quality Standard is under consideration.

Criteria Pollutants. The 1970 amendments to the Clean Air Act required the EPA to set National Ambient Air Quality Standards for certain pollutants known to be hazardous to human health. The EPA has identified and set standards to protect human health and welfare for six pollutants: ozone, carbon monoxide, total suspended particulates, sulfur dioxide, lead, and nitrogen oxide. The term "criteria pollutants" derives from the requirement that EPA must describe the characteristics and potential health and welfare effects of these pollutants in criteria documents. It is on the basis of these criteria documents that standards are set or revised.

Delegated State. A state (or other governmental entity) that has applied for and received authority to administer, within its territory, its state regulatory program as the federal program required under a particular federal statute.

Dose. The amount of a substance entering the body. Dose is calculated from the intake and the absorption efficiency, and it usually is expressed as mass of a substance absorbed into the body per unit body weight per unit time (e.g., mg/kg-day).

Emission Offsets. Compensation for a new or modified source of increased air emissions in the form of a binding commitment to reduce emission from another source.

Emission Reduction Banking. Setting aside or storing emission offsets for later use, trade, or sale.

Emissions Allowance. Under the 1990 CAA, the term used to represent the annual quantity of sulfur dioxide emissions (tons per year) from affected electric utility generating plants.

Emissions Trading. EPA policy that allows a plant complex with several facilities to decrease pollution from some facilities while increasing it from others, as long as total results are equal to or better than previous limits. Facilities where this is done are treated as if they exist in a bubble in which total emissions are averaged out. Complexes that reduce emissions substantially may "bank" their "credits" or sell them to other industries.

Fluid Modeling. Modeling conducted in a wind tunnel or water channel to quantitatively evaluate the influence of buildings and/or terrain on pollutant concentrations.

Fugitive Dust. Dust discharged to the atmosphere in an unconfined flow stream, such as that from unpaved roads, storage piles, and heavy construction operations.

Fugitive Emissions. Emissions that could not reasonably pass through a stack, chimney, vent, or other functionally equivalent opening (e.g., dust from unpaved roadways).

Gasoline Volatility. The property of gasoline whereby it evaporates into a vapor. Gasoline volatility is measured in pounds per square inch (psi), with a higher number reflecting more gasoline evaporation.

Halons. A family of compounds containing bromine used in fighting fires, whose breakdown in the atmosphere depletes stratospheric ozone.

Hazardous Air Pollutants. Air pollutants that are not covered by the criteria documents and the ambient air quality standards but which, as defined in the Clean Air Act, may reasonably be expected to cause or contribute to irreversible illness or death.

HCFCs. Chlorofluorocarbons that have been chemically altered by the addition of hydrogen. HCFCs are less reactive in the stratosphere than CFCs.

Increment. The amount of air pollution available in an attainment area, expressed in terms of pollutant concentration, for consumption by new or modified major sources.

Increment Consumption. The decrease in available pollution concentration owing to the construction of new or modified sources of emissions, or an increase in emissions at an existing source.

Lowest Achievable Emission Rate. Under the Clean Air Act, this is the rate of emission that reflects (a) the most stringent emission limitation that is contained in the implementation plan of any state for such source, unless the owner or operator of the proposed source demonstrates such limitations are not achievable; or (b) the most stringent emissions limitation achieved in practice, whichever is more stringent.

Low NO_x Burners. One of several combustion technologies used to reduce emissions of NO_x.

Maintenance Areas. Areas that recently attained the ambient air quality standards, but still have emission restrictions in place to ensure the area remains in attainment.

Major Modification. Any physical change in, or change in the method of operation of, a major stationary source that would result in a significant net emission increase of any regulated pollutant.

Major Stationary Source. Generally, any stationary facility or source of air pollution that directly emits, or has the potential to emit, pollutants in amounts greater than major source thresholds as defined by the legislation or regulations.

Material Safety Data Sheet (MSDS). A compilation of information required under the OSHA Communication Standard on the identity of hazardous chemicals, health, and physical hazards, exposure limits, and precautions. Section 311 of SARA requires facilities to submit MSDSs under certain circumstances.

Maximum Achievable Control Technology (MACT) Standards. Under the 1990 Amendments hazardous air pollutant program, the maximum degree of reduction in emissions achievable for new or existing sources, taking into consideration the cost of achieving such emissions reduction and any non-air-quality health and environmental impacts and energy requirements.

Mobile Source. A moving source of air pollution that can be licensed to travel on a public roadway, such as cars, trucks, and motorcycles. Mobile sources also include airplanes.

Montreal Protocol. An international environmental agreement to control chemicals that deplete the ozone layer. The protocol, which was renegotiated in June 1990, calls for a phaseout of CFCs, halons, and carbon tetrachloride by the year 2000, calls for a phaseout of chloroform by 2005, and provides financial assistance to help developing countries make the transition from ozone-depleting substances.

Netting. Process by which the preconstruction permit review is avoided through the use of the bubble concept by demonstrating that a contemplated modification will result in no overall net emissions increase within the major source.

Nonattainment Areas. Areas that have ambient air quality conditions worse than the national (or state or local) air quality standards.

Onboard Controls. Devices placed on vehicles to capture gasoline vapor during refueling and then route the vapors to the engine when the vehicle is started so that they can be efficiently burned (also termed Stage I controls).

Opacity. The amount of light obscured by particulate pollution in the air; clear window glass has a zero opacity, a brick wall has 100 percent opacity. Opacity is used as an indicator of changes in performance of particulate matter pollution control systems.

Operating Permit. The permit issued to a source after construction that allows the source to legally operate and emit pollutants. Operating permits often contain extensive conditions of operation that must be complied with.

Oxygenated Fuels. Gasoline that has been blended with alcohols or ethers that contain oxygen in order to reduce carbon monoxide and other emissions.

Ozone Depletion. Destruction of the stratospheric ozone layer that shields the earth from ultraviolet radiation harmful to biological life. This destruction of ozone is caused by the breakdown of certain chlorine- and/or bromine-containing compounds (chlorofluorocarbons or halons), which break down when they reach the stratosphere and catalytically destroy ozone molecules.

Point Source. A stationary location or fixed facility from which pollutants are discharged or emitted. Also, any single identifiable source of pollution—for example, a pipe, ditch, ship, ore pit, factory smokestack.

Potential to Emit. Emissions after controls and other federally enforceable limitations.

Preamble. The preamble to regulations promulgated by the EPA contains an explanation of and rationale for the regulations.

Precursor. In photochemical terminology, a compound such as a volatile organic compound (VOC) that "precedes" an oxidant. Precursors react in sunlight to form ozone or other photochemical oxidants.

Primary Standards. NAAQS reflective of the air quality necessary to protect the public health.

Reasonably Available Control Technology (RACT). The lowest emission limit that a particular source is capable of meeting by the application of control technology that is reasonably available, as well as technologically and economically feasible. RACT is usually applied to existing sources in nonattainment areas and in most cases is less stringent than new source performance standards.

Receptor. A location at which ambient air quality is measured or estimated.

Receptor Models. Procedures that examine an ambient monitor sample of particulate matter and the conditions of its collection to infer the types or relative mix of sources affecting it during collection.

Refined Model. An analytical technique that provides a detailed treatment of physical and chemical atmospheric processes and requires detailed and precise input data. Specialized estimates are calculated that are useful for evaluating source impact relative to air quality standards and allowable increments. The estimates are more accurate than those obtained from conservative screening techniques.

Reformulated Gasoline. Gasoline with a different composition from conventional gasoline (e.g., lower aromatics content) that results in the production of lower levels of air pollutants.

Residual Risk. The quantity of health risk remaining after application of MACT (Maximum Achievable Control Technology).

Risk Assessment. The qualitative and quantitative evaluation performed in an effort to define the risk posed to human health and/or the environment by the presence or potential presence and/or use of specific pollutants.

Risk Communication. The exchange of information about health or environmental risks between risk assessors, risk managers, the general public, news media, interest groups, and so on.

Risk Management. The process of evaluating alternative regulatory and nonregulatory responses to risk and selecting among them. The selection process necessarily requires the consideration of legal, economic, and social factors.

Rollback. A simple model that assumes that, if emissions from each source affecting a given receptor are decreased by the same percentage, ambient air quality concentrations decrease proportionately.

Sanctions. Actions taken against a state or local government by the federal government for failure to plan or to implement a SIP. Examples include withholding of highway funds and a ban on construction of new sources.

Screening Technique. A relatively simple analysis technique to determine if a given source is likely to pose a threat to air quality. Concentration estimates from screening techniques are conservatively high.

Secondary Standards. NAAQS reflective of the air quality necessary to protect public health.

Simple Terrain. An area where terrain features are all lower in elevation than the top of the stack of the source.

Stage II Controls. Systems placed on service station gasoline pumps to control and capture gasoline vapors during automobile refueling.

State Implementation Plan (SIP). Document prepared by states, and submitted to the EPA for approval, which identifies actions and programs to be undertaken by the state and its subdivisions to implement their responsibilities under the Clean Air Act.

Stationary Source. Any building, structure, facility, or installation that emits, or may emit, any air pollutant subject to regulations under the Clean Air Act. Stationary source emissions include emissions from moving but not "mobile" sources, such as forklifts.

Stratosphere. That portion of the atmosphere above about ten miles from the Earth's surface. The stratosphere is not directly related to weather at the surface (see "Troposphere"), but provides radiation protection to humans.

Top-Down BACT. An approach to determining appropriate BACT for a PSD source under which the applicant must determine the most stringent control available for a similar or identical source or source category. Only if it can be shown that this level of control is technically or economically infeasible for the source in question will the next most stringent level of control be determined, similarly evaluated, and incorporated in the PSD permit.

Transportation Control Measures (TCMs). Steps taken by a locality to adjust traffic patterns (e.g., bus lanes, right turn on red) or reduce vehicle use (ride-sharing, high-occupancy vehicle lanes) to reduce vehicular emissions of air pollutants.

Troposphere. The lower atmosphere, that portion within about ten miles of the Earth's surface. The troposphere contains the weather (clouds, wind, and precipitation).

Vehicle Miles Traveled (VMT). A measure of both the volume and extent of motor vehicle operation; the total number of vehicle miles traveled within a specified geographical area (whether the entire country or a smaller area) over a given period of time.

APPENDIX

C

1990 CAA AMENDMENT PROVISIONS RELATED TO SPECIFIC STATIONARY SOURCE CATEGORIES

Note to the Reader: This is a reprint of a document prepared by the USEPA OAQPS (1993). Although the USEPA reviewed the 1990 Clean Air Act Amendments in order to compile this listing of provisions related to specific stationary source categories, this listing may not represent the universe of source categories which are addressed by the Amendments.

INTRODUCTION

The 1990 Clean Air Act Amendments (CAAA) contain a number of provisions that pertain to specific stationary source categories. These source categories representing both anthropogenic and natural sources include:

- Aerospace
- Boat manufacturing
- Coke ovens
- Consumer and commercial products
- Electric utilities
- Gasoline dispensing facilities
- Hazardous waste treatment/storage disposal facilities
- Marine vessels
- Medical and research facilities
- Oil and gas wells
- Outer continental shelf sources
- Publicly owned treatment works
- Rocket engines and motor firing
- Sources which emit methane
- Sources which emit radionuclides
- Solid/municipal waste combustion
- Surface coal mines

Summaries of the provisions affecting these source categories are presented by Title and section number in the following paragraphs.

AEROSPACE

Title I

Control Technique Guidelines (CTG) Section 183 (b) (3) requires the EPA, within 3 years of enactment of the CAAA, to issue control technique guidelines (CTGs) for reducing emissions of volatile organic compounds (VOCs) from the aerospace coatings and solvents industry. The CTGs shall specify the best achievable control measures (BACM) to use in reducing the emissions of VOCs from coating operations and solvent usage in the aerospace industry. The EPA shall consult with the Departments of Defense and Transportation and the National Aeronautics and Space Administration in developing these CTGs.

BOAT MANUFACTURING

Title III

List of Source Categories Section 112 (c) (8) requires the EPA to list boat manufacturing as a separate subcategory when establishing emission standards for styrene, unless the EPA determines that such a listing is inconsistent with the goals and requirements of the Clean Air Act (CAA).

COKE OVENS

Title III

Petitions to Modify the Hazardous Pollutant List Section 112 (b)(3) allows persons to petition the EPA to remove certain unique substances from the pollutants listed in Section 112 (b)(1) which do not have CAS numbers, with the exception of *coke oven emissions*, mineral fibers, and polycyclic organic matter.

Emissions Standards Section 112 (d)(8)(A) requires the EPA to promulgate technology-based emissions standards for coke oven batteries by December 31, 1992. The regulations shall require that coke oven batteries not exceed 8 per centum leaking doors, 1 per centum leaking lids, 5 per centum leaking offtakes, and 16 seconds visible emissions per charge with no exclusion for emissions during the period after the closing of self-sealing doors. The compliance date for existing coke oven batteries for meeting the emissions standards is December 31, 1995. Section 112 (d)(8)(B) requires the EPA to promulgate regulations for work practices applicable to coke oven batteries. The compliance date for these regulations will be 3 years from the enactment date of the 1990 CAAA.

Extensions from the Health Based Standards Sections 112 (d)(8)(C) and 112 (i)(8) extend the compliance date for coke ovens until January 1, 2020, for the health based standards promulgated under Section 112 (f). Coke oven batteries which elect to qualify for such compliance extension must comply with the emissions standards promulgated under Section 112 (d)(8)(A) as well as comply with other requirements specified in Section 112 (i)(8)(A), (B), (C), and (E).

Coke Oven Production Technology Study Section 112 (n)(2) requires the EPA and the Department of Energy (DOE) to undertake a 6-year study of coke oven production emission control technologies in order to develop and commercialize such technologies which may potentially significantly reduce emissions of hazardous air pollutants from coke oven production facilities.

CONSUMER OR COMMERCIAL PRODUCTS

Title I

Study and Report Section 183 (e)(2) requires the EPA to submit a report to Congress within 3 years of enactment of the CAAA on the results of a study of the emissions of VOCs from commercial and consumer products.

Regulations Section 183 (e) (3) requires the EPA to list categories of consumer or commercial products that account for at least 80 percent of the VOC emissions (reactivity-adjusted) in areas violating the NAAQS for ozone. These listed categories are required to be divided into four groups establishing priorities for regulations. EPA must promulgate regulations requiring best available controls for each group every 2 years (beginning after the list is promulgated) until all four groups are regulated.

ELECTRIC UTILITIES

Title I

SIP Revisions Section 182 (e) (3) requires states which contain all or part of an extreme ozone nonattainment area to submit SIP revisions, within

3 years after enactment, that require new, modified and existing electric utilities and industrial and commercial boiler units which emit more than 25 tons per year of NO_x to (1) burn natural gas, methanol, or ethanol (or a comparably low polluting fuel) as the primary fuel and (2) use advanced control technologies (such as catalytic control technology or other comparable control methods) to reduce emissions of oxides of nitrogen.

Title III

EPA Report on Health Hazards Section 112 (n) (1) (A) requires the EPA to report to Congress, within 3 years after enactment, the results of a study on the public health hazards that are reasonably anticipated to occur as a result of emissions of listed hazardous air pollutants from electric utility steam generating units after imposition of CAA requirements. The report must include alternative control strategies for emissions that may warrant regulation. The EPA must regulate electric utility steam-generating units if the Agency determines that it is appropriate and necessary based on the results of the health effects assessment study.

Mercury Emissions Study Section 112 (n) (1) (B) and (C) requires the EPA to conduct and transmit to Congress no later than 4 years after enactment of the CAAA a study of mercury emissions from electric utility steam-generating units and other sources, including area sources.

The National Institute for Environmental Health Sciences (NIEHS) shall conduct a study, within 3 years of enactment, to determine the threshold effect level of mercury exposure below which adverse human health effects are not expected to occur.

Other Categories of Radionuclide Emissions

Section 112 (q)(3) stipulates that requirements in effect prior to enactment of the CAAA for radionuclide emissions from (1) non-DOE federal facilities that are not licensed by the NRC, (2) coal-fired utility and industrial boilers, (3) underground uranium mines, (4) surface uranium mines, and (5) disposal of uranium mill tailings piles shall remain in effect unless the EPA applies to these radionuclide sources the requirements as modified by the CAAA.

Title IV

SO_2 Allowance Program In general, all Title IV provisions affect electric utilities. Title IV establishes a two-phase program for reducing sulfur dioxide (SO_2) emissions by 10 million tons from 1980 emission levels. Phase I emission reductions begin in 1995 and the Phase II reduction program will begin in the year 2000. For each phase of the reduction program, sources will receive "allowances" equal to the number of tons of sulfur dioxide it is permitted to emit. One allowance is defined as the authorization to emit one ton of SO_2. In general, electric utilities will be given allowances in accordance with the following table:

Phase	Date	Allowance Level	Applicability
I	1/1/1995	2.5 lbs SO_2/mmBTU	111 Listed Power Plants
II	1/1/2000	1.2 lbs SO_2/mmBTU	All Plants

Sources which reduce their emissions by more than the required emissions levels can sell their extra allowance(s) to other sources which will allow these sources to increase their emissions while remaining in compliance. Because this two-phase program will affect primarily utility power plants, industrial facilities with boilers over a certain size which use more than one-third of their BTU output to produce electricity may also be affected. Furthermore, the EPA is required to conduct studies of SO_2 emissions from industrial facilities and take the appropriate steps to ensure that emissions from such sources do not exceed 5.6 million tons per year (TPY). Subsequent reports to Congress of such studies are required every 5 years.

During the Phase I reduction program, Section 404 allows for additional allowances to be given to certain sources listed in Illinois, Indiana, and Ohio and defines the criteria for other utilities to qualify for these additional allowances.

Section 405 allows for bonus allowances for the following four groups of electric utilities during the Phase II program: (1) existing utility units which serve generators with a capacity equal to or above 75 megawatts and with an actual 1985 emission rate equal to or greater than 1.2 lb/mmBTU; (2) coal or oil-fired existing utility units which serve generators of less than 75 megawatts and with actual 1985 emission rates equal to or greater than 1.2 lb/mmBTU; (3) existing coal-fired utility units for which the lesser of actual or allowable 1985 emission rates were less than 1.2 lb/mmBTU; and (4) oil and gas-fired units whose annual average fuel consumption during 1980 to 1989 exceeded 90 percent in the form of natural gas. In lieu of these bonus allowances, "clean" states may elect to have allocated special bonus allowances under Section 406 for all fossil fuel-fired utility steam generating units.

NO_x Reduction Program Title IV also includes a program for reducing emissions of nitrous oxides (NO_x) from utility boilers by 2 million tons from the year 2000 projected levels. Section 407 requires the EPA to promulgate regulations for setting emission standards for utility boilers which are to be applicable to tangentially fired and dry-bottom wall-fired boilers by 1995. Title IV sets maximum allowable emission rates above which the standards cannot be set unless the EPA determines that such emission rates cannot be achieved using low NO_x burner technology. By 1997, the EPA must conduct a study of the feasibility of trading SO_2 allowances for NO_x allowances.

Other Requirements Additional Title IV requirements for electric utilities include (1) installing continuous emission monitors (CEMs) or an EPA-defined alternate monitoring system, (2) developing compliance plans, (3) implementing quality assurance procedures for emission data systems, (4) assessing annual fees of $2000 per ton for emissions in excess of allowance levels, and (5) provisions for clean coal technology demonstration projects. Title IV also requires the EPA to undertake additional studies by November 1993 to determine if additional controls are necessary to control acidic deposition.

Title IX

Pollution Prevention and Emission Control Research Section 901 (c) of the CAAA amends Section 103 (g) of the CAA by requiring the EPA to conduct a research program to develop, evaluate, and demonstrate nonregulatory strategies and technologies for air pollution prevention for air pollutants which pose a significant human health and environmental risk. The research program must include improvements in nonregulatory strategies and techniques for preventing or reducing multiple air pollutants including SO_x, NO_x, heavy metals, PM_{10}, CO, and CO_2 from stationary sources including *fossil fuel power plants.*

GASOLINE DISPENSING FACILITIES

Title I

SIP Provisions Section 182 (b)(3) requires states which contain all or part of a moderate ozone nonattainment area(s) to submit to the EPA, within 2 years after enactment, a SIP revision which requires gasoline dispensing facilities (which sell more than 10,000 gallons per month or 50,000 gallons per month in the case of an independent small business marketer of gasoline) to install and operate a gasoline vapor recovery system for recovering emissions caused from the refueling of motor vehicles.

HAZARDOUS WASTE TREATMENT/ STORAGE-DISPOSAL FACILITIES

Title I

Control Technique Guidelines (CTG) Section 183 (b)(2) requires the EPA to give priority, in developing CTGs, to those source categories which are the most significant contributors to ozone air pollution in ozone nonattainment areas. Such categories shall include *hazardous waste treatment, storage, and disposal facilities* (*TSDF*) permitted under subtitle C of the Solid Waste Disposal Act.

Title III

Clean Air Act Versus RCRA Regulations Section 112 (n)(7) requires the EPA to ensure to the maximum extent practicable that the requirements pertaining to any category or subcategory of sources of air emissions regulated under subtitle C of the Solid Waste Disposal Act [e.g., *hazardous waste treatment, storage, and disposal facilities* (*TSDF*)] and Title III are consistent.

MARINE VESSELS

Title I

Control Technique Guidelines (CTG) Section 183 (b)(4) requires the EPA, within 3 years of enactment of the CAAA, to issue control technique guidelines (CTGs) for reducing emissions of volatile organic compounds (VOCs) and PM_{10} from painting, coating, and solvents operations in the shipbuilding and ship repair industry. The guidelines shall specify the best achievable control measures (BACM) to use in reducing the emissions of VOCs and PM_{10} from operations at these industries. The EPA is also required to consult with the appropriate Federal agencies in developing these CTGs.

Emission Standards Section 183 (f)(1) requires the EPA, within 2 years after enactment, to promulgate standards applicable to VOC emissions and any other pollutant from the loading and unloading of marine tank vessels that may reasonably be anticipated to endanger public health or welfare. The standards would require *reasonably available control technology* (RACT), considering costs, any non-air-quality benefits, environmental impacts, energy requirements, and safety factors associated with alternative control techniques.

Equipment Safety Section 183 (f) (2) requires the U.S. Coast Guard, within 6 months after enactment, to issue regulations for ensuring the safety of equipment and operations which are to control emissions from the loading and unloading of tank vessels. Standards and regulations promulgated by EPA and state agencies shall be consistent with regulations promulgated by the Coast Guard.

Title VIII

Contaminated Used Oil in Ships Section 813 requires the EPA (in consultation with the Secretary of Transportation and Secretary of the Department in which the Coast Guard is operating), within 2 years of enactment, to submit a report to Congress which evaluates the health and environmental impacts of the combustion of contaminated used oil in ships, the reason for using such oil, the alternatives to such use, and the costs of such alternatives.

MEDICAL AND RESEARCH FACILITIES

Title III

Specific Source Categories Section 112 (c) requires the EPA, in publishing the list of categories of major and area sources, to establish a separate category for research or laboratory facilities to assure equitable treatment of these facilities.

Emissions Standards for Medical Facilities Section 112 (q)(4) stipulates that emissions standards promulgated prior to enactment of the CAAA for medical research or treatment facilities shall not take effect until 2 years after enactment, unless the EPA determines otherwise under Section 112 (d)(9).

OIL AND GAS WELLS

Title III

Source Categorization Section 112 (n) stipulates that emissions from any oil or gas exploration or production well and pipeline compressor or pump station may not be aggregated with emissions from other similar units for purposes of determining if such sources are major sources. Furthermore, oil and gas production wells may not be listed as an area source category, except in the case where emissions from oil and gas production wells located in a Metropolitan Statistical Area (MSA) or a Consolidated Metropolitan Statistical Area (CMSA) with a population of 1 million or more are determined to present more than a negligible risk of adverse effects to public health.

Hydrogen Sulfide Emissions Assessments Section 112 (n)(5) requires the EPA to assess the hazards to public health and environment resulting from hydrogen sulfide emissions from the extraction of oil and natural gas. The results of this study must be submitted to Congress within 24 months after enactment.

Title VI

Domestic Methane Source Inventory and Control Section 603 (b) requires the EPA, DOE, and Department of Agriculture to study methane emissions (including an inventory of methane generation) from the following: (1) natural gas operations which shall include but are not limited to, accidental and intentional releases from *natural gas and oil wells*, pipelines, processing facilities, and gas burners; (2) coal extraction processes which shall include, but are not limited to, accidental and intentional releases from mining shafts, degasification wells, gas recovery wells and equipment, and coal processing and use operations (including an inventory of methane generation); (3) solid waste operations which shall include an inventory of emissions from waste management operations, including storage, treatment and disposal; (4) agricultural operations to include an inventory of emissions from rice and livestock production in the United States; (5) emissions from biomass burning to include an inventory of emissions from the intentional burning of agricultural wastes, wood, grasslands and forests; and (6) other methane emissions from human activities deemed significant by the EPA and other agencies.

Title VIII

Exemptions for Stripper Wells Section 819 stipulates that ozone, CO, PM_{10}, SO_2, NO_2, and lead nonattainment provisions contained in Title I shall not apply with respect to the production of and equipment used in the exploration, production, development, storage, or processing of oil from stripper well property or stripper well natural gas except that the Title I provisions cover (1) serious nonattainment areas having a population of 350,000 or more or (2) severe or extreme nonattainment areas.

OUTER CONTINENTAL SHELF (OCS) SOURCES

Title VIII

Regulations Section 801 of the CAAA adds Section 328 to the CAA which requires the EPA (following consultation with the Secretary of the Interior and the Commandant of the U.S. Coast Guard), within 12 months after enactment, to establish requirements (including emissions controls, emission limitations, offsets, permitting, monitoring, testing, and reporting) to control air pollution from OCS sources to attain and maintain Federal and State ambient air quality standards and comply with Prevention of Significant Deterioration (PSD) provisions. These requirements are applicable to OCS sources located offshore of the states along the Pacific, Arctic, and Atlantic Coast, and the U.S. Gulf Coast off the State of Florida. New OCS sources are required to comply with the requirements on the date of promulgation and existing sources must comply within 24 months after promulgation. The EPA may exempt an OCS source from a control technology requirement if it is technically infeasible or would cause an unreasonable threat to health and safety. If a state is located adjacent to an OCS source, the state may promulgate regulations to implement and enforce the requirements of this subsection. If the EPA determines that the regulations are adequate, the EPA may grant authority to the state to implement and enforce these regulations.

Requirements for Other Offshore Areas and Research Section 801 of the CAAA adds Section 328 (b) to the CAA which requires that the EPA and Coast Guard assure the coordination of regulations for OCS emissions off the coast of Texas, Louisiana, Mississippi, and Alabama. Within 3 years of enactment, the Coast Guard must conduct a research study examining the impact of emissions from OCS activities in areas which fail to meet the ozone or NO_2 NAAQS.

PUBLICLY OWNED TREATMENT WORKS (POTW)

Title III

Emission Standards Section 112 (e)(5) requires the EPA to promulgate technology-based (MACT) emission standards for POTWs within 5 years after enactment of the CAAA.

Hazardous Air Pollutant Emissions Studies Section 112 (n)(3) authorizes the EPA to conduct studies in cooperation with the owners and operators of POTWs to characterize hazardous air pollutant emissions, identify industrial, commercial, and residential discharges that contribute to such emissions, and demonstrate control measures for such emissions.

ROCKET ENGINE AND MOTOR FIRING SOURCES

Title I

Emission Offsets Section 173 (e) requires states which permit rocket engine and motor firing facilities to allow emissions offsets from new and existing facilities provided that (1) facility modifications are for the sole purpose of testing and firing rocket engines or motors, (2) the source has used all reasonable means to obtain and utilize emissions offsets, (3) the DOE, DOT, NASA and other appropriate Federal agencies have determined that the rocket testing at such facilities is essential to the national security, and (4) the source complies with other alternative offset measures imposed by the permitting agency.

SOURCES WHICH EMIT METHANE

Title VI

Methane Studies Section 603 requires the EPA, within 2 years after enactment, to prepare and submit to Congress a report that identifies activities, substances, processes, or combinations thereof that could reduce emissions of methane and which are economically and technologically justified with and without consideration of environmental benefit. The study (to be conducted in coordination with the DOE and Department of Agriculture) is to examine methane emissions (including an inventory of methane generation) from the following: (1) natural gas operations which shall include, but are not limited to, accidental and intentional releases from natural gas and oil wells, pipelines, processing facilities, and gas burners; (2) coal extraction processes which shall include, but are not limited to, accidental and intentional releases from mining shafts, degasification wells, gas recovery wells and equipment, and coal processing and use operations (including an inventory of methane generation); (3) solid waste operations which shall include an inventory of emissions from waste management operations, including storage, treatment, and disposal; (4) agricultural operations to include an inventory of emissions from rice and livestock production in the United States; (5) emissions from biomass burning to include an inventory of emissions from the intentional burning of agricultural wastes, wood, grasslands, and forests; and (6) other methane emissions from human activities deemed significant by the EPA and other agencies. In addition, the study shall examine methane emissions from *biogenic sources* such as tropical, temperate and subarctic forests, tundra, and freshwater and saltwater wetlands.

SOURCES WHICH EMIT RADIONUCLIDES

Title III

Emissions Standards Section 112 (d)(9) stipulates that the EPA shall not promulgate standards for radionuclide emissions from a source category that is licensed by the NRC if the EPA determines (by rule and after consultation with the NRC) that the NRC regulatory program provides an ample margin of safety to protect public health.

Special Rule Section 112 (q)(2) stipulates that the EPA shall not promulgate standards for radionuclide emissions from elemental phosphorous plants, phosphogypsum stacks, or grate calcination elemental phosphorous plants.

Other Categories of Radionuclide Emissions Section 112 (q) (3) stipulates that requirements in effect prior to enactment of the CAAA for radionuclide emissions from (1) non-DOE federal facilities that are not licensed by the NRC, (2) coal-fired utility and industrial boilers, (3) underground uranium mines, (4) surface uranium mines, and (5) disposal of uranium mill tailings piles shall remain in effect unless the EPA applies to these radionuclide sources the requirements as modified by the CAAA.

Accident Prevention Section (r)(7)(D) stipulates that emissions of radionuclides from the construction or operation of sources regulated by the NRC are not subject to the accidental release prevention regulations which the EPA is required to promulgate.

SOLID/MUNICIPAL WASTE COMBUSTION

Title III

Mercury Emissions Study Section 112 (n)(1) (B) and (C) requires the EPA to conduct and transmit to Congress, no later than 4 years after enactment of the CAAA, a study of mercury emissions from municipal waste combustion units and other sources, including area sources.

NIEHS shall conduct a study within 3 years of enactment to determine the threshold effect level of mercury exposure below which adverse human health effects are not expected to occur.

New Source Performance Standards Section 129 (a) requires the EPA to establish performance standards and other requirements (pursuant to the NSPS requirements) for each category of *solid waste incineration units,* including emissions limits and other requirements for new units, and guidelines and other requirements for existing units. For solid waste incineration units with capacity *greater than 250 tons per day* combusting *municipal waste,* standards must be promulgated no later than *12 months* after enactment. For such units with a capacity equal to or *less than 250 tons per day* and units combusting *hospital waste* (i.e., medical and infectious waste), the standards must be promulgated no later than *24 months* after enactment. For other categories of solid waste incineration units, the EPA must publish a schedule for promulgation of standards no later than *18 months* after enactment.

The emission standards must reflect the *maximum degree of emissions reduction,* considering cost and any non-air-quality health and environmental impacts and energy requirements deemed achievable for new or existing units in each category. The EPA may distinguish between classes, types, and sizes of units within a category.

For new units, the emission reduction deemed achievable must not be less stringent than the control achieved in practice by the *best controlled similar unit* (as determined by the EPA).

For solid waste incineration units, the standards must be based on *removal or destruction* technologies before, during, or after combustion and must include siting requirements for new units to minimize (to the maximum extent practicable) potential risks to human health or the environment, on a site-specific basis.

Numerical emissions limits must be established for particulate matter, opacity, sulfur dioxide, hydrogen chloride, NO_x, CO, lead, cadmium, mercury, dioxins, and dibenzofurans. The EPA may promulgate numerical emissions limits for other pollutants as well.

Existing Solid Waste Incineration Units Section 129 (b) requires the EPA to establish *guidelines* for existing units, including emission limits, monitoring, operator training, and permit requirements as specified. *For existing units*, the emissions standards may be less stringent than for new units in the same category, but not less stringent than the average emissions limit achieved by the *best performing 12 percent of units* in the category (excluding units that first met the lowest achievable emissions rates 18 months prior to proposal of the standards, or 30 months prior to the date the standards are promulgated, whichever is later).

States are required to submit, within 1 year after promulgation of the guidelines, plans to implement and enforce the guidelines, providing for compliance no later than 3 years after approval of the State plan but also no later than 5 years after promulgation of the guidelines.

The EPA has *180 days* to approve or disapprove the plan; reasons for disapproval must be put in writing.

The EPA must develop, implement, and enforce a *Federal plan* for existing units if the state has not submitted an approvable plan within *2 years* after promulgation of the guidelines. The Federal plan must ensure compliance no later than *5 years* after promulgation of the guidelines.

Monitoring Section 129(c) requires the EPA, as part of the performance standards, to promulgate regulations requiring owners or operators of each solid waste incineration unit to monitor emissions and report findings. The regulations must include requirements on the frequency of monitoring, test methods and procedures, and the form and frequency of reports.

Operator Training Section 129(d) requires the EPA, within 24 months after enactment, to establish a model State training and certification program for solid waste incineration unit operators and high-capacity fossil fuel-fired plant operators. This section also stipulates that it will be *unlawful* for any person having control over processes affecting emissions from applicable solid waste incineration units and high-capacity fossil fuel-fired plants to operate such units beginning *36 months* after promulgation of the performance standards and guidelines, unless the person has satisfactorily completed an approved training program.

Permit Requirements Section 129(e) stipulates that, beginning *36 months* after promulgation of the performance standard for a solid waste incineration unit (or the effective date of the applicable State program, whichever is later), all units must have a permit issued under Title V. Permits for solid waste incineration units combusting municipal waste are to be issued for a period up to *12 years* and must be reviewed every *5 years*.

Residual Risk Section 129 (h)(3) requires the EPA to promulgate health-based standards if such standards are required by Section 112 (f).

SURFACE COAL MINES

Title II

Emission Factors Section 234 requires the EPA to review and revise, as necessary, AP-42 emission factors used for modeling fugitive particulate emissions from surface coal mines for demonstrating compliance with the NAAQS or for purposes of new source review.

APPENDIX

D

ATMOSPHERIC DISPERSION COEFFICIENTS

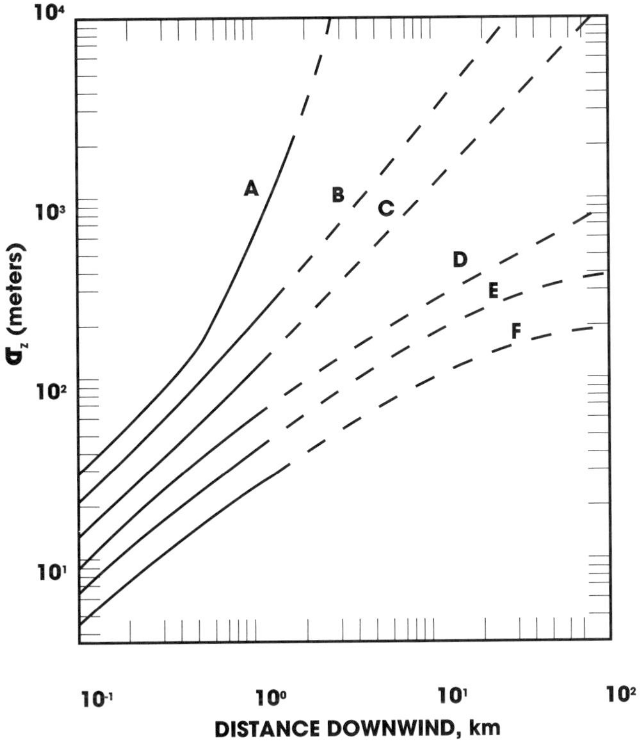

■ **FIGURE D-1** Vertical Dispersion Coefficients. *Source:* After Hanna et al. (1982).

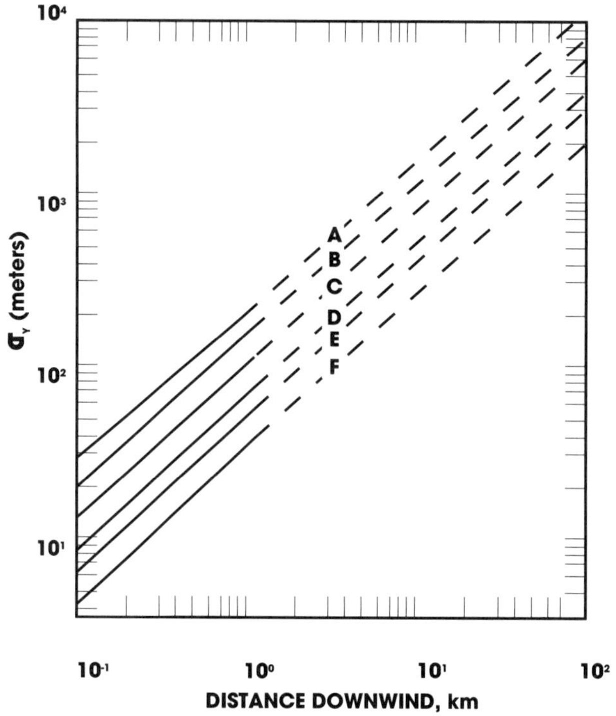

■ **FIGURE D-2 Horizontal Dispersion Coefficients.** *Source:* After Hanna et al. (1982).

APPENDIX

E

DOSE–RESPONSE VALUES FOR SELECTED AIR TOXICS

■ **TABLE E-1** Dose–Response Values for Selected Air Toxics[a]

Substance	Cancer Potency Value Unit Risk	Noncancer Chronic AEL (μg/m³)	Noncancer Acute AEL (μg/m³)	Multi-pathway Exposure/ Oral Potency (mg/kg-d)⁻¹
Acetaldehyde	2.2E – 6	9.0E + 0		
Acetamide	2.0E – 5			
Acetonitrile				
Acetophenone				
2-Acetylaminofluorene	9.4E – 6			
Acrolein		2.0E – 2	2.5E + 0	
Acrylamide	1.3E – 3	7.0E – 1		
Acrylic acid				
Acrylonitrile	2.9E – 4	2.0E + 0		
Allyl chloride				
4-Aminobiphenyl				
Aniline				
o-Anisidine				
Asbestos	(1.9E – 4/100 fibers/m³)			
Benzene (including benzene from gasoline)	2.9E – 5			
Benzidine	1.4E – 1			
Benzotrichloride				
Benzyl chloride		1.2E + 1	5.0E + 1	
Biphenyl				
Bis(2-ethylhesyl) phthalate (DEHP)				
Bis(chloromethyl)ether	1.3E – 2			
Bromoform				
1,3-Butadiene	2.8E – 4			
Calcium cyanamide				
Caprolactam				
Captan				
Carbaryl				

■ **TABLE E-1** Dose–Response Values for Selected Air Toxics (*Continued*)

Substance	Cancer Potency Value Unit Risk	Noncancer Chronic AEL (μg/m³)	Noncancer Acute AEL (μg/m³)	Multi-pathway Exposure/ Oral Potency (mg/kg-d)⁻¹
Carbon disulfide				
Carbon tetrachloride	4.2E − 5	2.4E + 0	1.9E + 2	
Carbonyl sulfide				
Catecol				
Chloramben				
Chlordane				
Chlorine		7.1E + 0	2.3E + 1	
Chloroacetic acid				
2-Chloroacetophenone				
Chlorobenzene		7.0E + 1		Yes/NA[b]
Chlorobenzilate				
Chloroform	5.3E − 6	3.5E + 1		
Chloromethyl methyl ether				
Chloroprene	1.3E − 7	1.0E + 0		
Cresols/cresylic acid (isomers and mixture)				
o-Cresol		1.8E + 2		
m-Cresol		1.8E + 2		
p-Cresol		1.8E + 2		
Cumene				
2,4-D (2,4-dichloro-phenoxyacetic acid) salts and esters				
DDE				
Diazomethane				
Dibenzofurans		3.5E − 6		Yes/1.33E + 5
1,2-Dibromo-3-chloro-propane	2.0E − 3	2.0E − 1		
Dibutylphthalate				

■ **TABLE E-1** *(Continued)*

Substance	Cancer Potency Value Unit Risk	Noncancer Chronic AEL (µg/m³)	Noncancer Acute AEL (µg/m³)	Multi-pathway Exposure/ Oral Potency (mg/kg-d)⁻¹
1,4-Dichlorobenzene (*p*)	1.1E − 5	7.0E + 2		Yes/0.04
3,3-Dichlorobenzidine	3.4E − 4			
Dichloroethyl ether (bis(2-chloroethyl) ether)				
1,3-Dichloropropene				
Dichlorvos				
Diethanolamine				
N,N-Diethyl aniline (*N,N*-dimethylaniline)				
Diethyl sulfate				
3,3-Dimethoxybenzidine				
3,3′-Dimethyl benzidine				
Dimethyl carbomoyl chloride				
Dimethyl formamide				
1,1-Dimethyl hydrazine	4.9E − 4			
Dimethyl phthalate				
Dimethyl sulfate	4.0E − 3			
4,6-Dinitro-*o*-cresol and salts				
2,4-Dinitrophenol				
2,4-Dinitrotoluene				
1,4-Dioxane (1,4-diethyleneoxide)	7.7E − 6	4.0E + 0	2.0E + 1	
1,2-Diphenylhydrazine				
Epichlorohydrin (1-chloro-2, 3-epoxypropane)	2.3E − 5	3.0E − 1		
1,2-Epoxybutane				
Ethyl acrylate		4.8E + 1		

■ **TABLE E-1** Dose–Response Values for Selected Air Toxics (*Continued*)

Substance	Cancer Potency Value Unit Risk	Noncancer Chronic AEL (μg/m³)	Noncancer Acute AEL (μg/m³)	Multi-pathway Exposure/ Oral Potency $(mg/kg\text{-}d)^{-1}$
Ethyl benzene				
Ethyl carbamate (urethane)				
Ethyl chloride (chloroethane)		1.0E + 4		
Ethylene dibromide (dibromoethane)	7.1E – 5	4.6E + 0		
Ethylene dichloride (1,2-dichloroethane)	2.0E – 5	9.5E + 1		
Ethylene glycol				
Ethylene imine (aziridine)				
Ethylene oxide	8.8E – 5	6.0E + 2		
Ethylene thiourea	1.3E – 5			
Ethylidene dichloride (1,1-dichloroethane)				
Formaldehyde	1.3E – 5	3.6E + 0	3.7E + 2	
Heptachlor				
Hexachlorobenzene	5.1E – 4	2.8E + 0		Yes/1.8
Hexachlorobutadiene				
Hexachlorocyclopent-adiene		2.4E – 1		
Hexachloroethane				
Hexamethylene-1, 6-diisocyanate				
Hexamethylpho-sphoramide				
Hexane				
Hydrazine	4.9E – 3	2.4E – 1		
Hydrochloric acid [hydrogen chloride (gas only)]		7.0E + 0	3.0E + 3	

■ **TABLE E-1** (*Continued*)

Substance	Cancer Potency Value Unit Risk	Noncancer Chronic AEL (μg/m³)	Noncancer Acute AEL (μg/m³)	Multi-pathway Exposure/ Oral Potency (mg/kg-d)⁻¹
Hydrogen fluoride (hydrofluoric acid)		5.9E + 0	5.8E + 2	
Hydroquinone				
Isophorone				
Lindane				
Maleic anhydride		2.4E + 0	1.0E + 1	
Methanol		6.2E + 2		
Methoxychlor				
Methyl bromide (bromomethane)		6.0E + 0		
Methyl chloride (chloromethane)				
Methyl chloroform (1,1,1-trichloroethane)		3.2E + 2	1.9E + 5	
Methyl ethyl ketone (2-butanone)				
Methyl hydrazine				
Methyl iodide (iodomethane)				
Methyl isobutyl ketone (hexone)				
Methyl isocyanate		3.6E – 1		
Methyl methacrylate		9.8E + 2		
Methyl tert-butyl ether				
4,4′-Methylenebis (2-chloroaniline)	3.7E – 5			
Methylene chloride (dichloromethane)	1.0E – 6	3.0E + 3	3.5E + 3	
Methylene diphenyl diisocyanate (MDI)				
4,4-Methylenedianiline	3.4E – 4	1.9E + 0		
Naphthalene		1.4E + 1		

■ TABLE E-1 Dose–Response Values for Selected Air Toxics (*Continued*)

Substance	Cancer Potency Value Unit Risk	Noncancer Chronic AEL (μg/m³)	Noncancer Acute AEL (μg/m³)	Multi-pathway Exposure/ Oral Potency $(mg/kg\text{-}d)^{-1}$
Nitrosobenzene				
4-Nitrobiphenyl				
4-Nitrophenol				
2-Nitropropane	1.3E – 3	2.0E + 1		
N-Nitroso-*N*-methylurea				
N-Nitroso-dimethylamine	4.6E – 3			Yes/36
N-Nitrosomorpholine	1.9E – 3			Yes/6.7
Parathion				
Pentachloronitro-benzene				
Pentachlorophenol	4.6E – 6	2.0E – 1		Yes/0.016
Phenol		4.5E + 1		
p-Phenylenediamine				
Phosgene			1.2E + 1	
Phosphine		1.0E + 1		
Phosphorus		7.0E – 2		
Phthalic anhydride		7.0E + 3		
Polychlorinated biphenyls (Aroclors)	1.4E – 3	1.2E + 0		Yes/7.7
1,3-Propane sultone	6.9E – 4			
beta-Propiolactone				
Propionaldehyde				
Proposur (Baygon)				
Propylene dichloride (1,2-dichloropropane)				
Propylene oxide	3.7E – 6	3.0E + 1	1.0E + 3	
1,2-Propylenimine (2-methyl aziridine)				
Quinoline				

■ **TABLE E-1** ***(Continued)***

Substance	Cancer Potency Value Unit Risk	Noncancer Chronic AEL (μg/m³)	Noncancer Acute AEL (μg/m³)	Multi-pathway Exposure/ Oral Potency (mg/kg-d)$^{-1}$
Quinone				
Styrene	5.7E − 7	7.0E + 2		
Styrene oxide				
2,3,7,8-Tetrachloro-dibenzo-*p*-dioxin		3.5E − 6		Yes/1.33E+5
1,1,2,2-Tetrachloro-ethane				
Tetrachloroethylene (perchloroethylene)	5.8E − 7	3.5E + 1	6.8E + 3	
Titanium tetrachloride				
Toluene		2.0E + 2		
2,4-Toluene diamine				
2,4-Toluene diisocyanate	1.1E − 5	9.5E − 2		
o-Toluidine				
Toxaphene (chlorinated camphene)				
1,2,4-Trichlorobenzene				
1,1,2-Trichloroethane		6.4E + 2		
Trichloroethylene	2.0E − 6			
2,4,5-Trichlorophenol				Yes/NA
2,4,6-Trichlorophenol	2.0E − 5			Yes/0.07
Triethylamine				
Trifluralin				
2,2,4-Trimethylpentane				
Vinyl acetate				
Vinyl bromide				
Vinyl chloride	7.8E − 5	2.6E + 1		
Vinylidene chloride (dichloroethylene)		3.2E + 1		

■ **TABLE E-1** Dose–Response Values for Selected Air Toxics (*Continued*)

Substance	Cancer Potency Value Unit Risk	Noncancer Chronic AEL (μg/m³)	Noncancer Acute AEL (μg/m³)	Multi-pathway Exposure/ Oral Potency (mg/kg-d)⁻¹
Xylenes (isomers and mixture)		3.0E + 2	4.4E + 3	
o-Xylenes				
m-Xylenes				
p-Xylenes				
Antimony compounds				
Arsenic compounds (inorganic including arsine)				
Arsenic	3.3E – 3	5.0E – 1		Yes/1.7
Beryllium compounds				
Beryllium	2.4E – 3	4.8E – 3		Yes/4.3
Cadmium compounds				
Cadmium	4.2E – 3	3.5E + 0		Yes/NA
Chromium compounds				
Hexavalent chromium	1.4E – 1	2.0E – 3		Yes/0.42
Cobalt compounds				
Coke oven emissions	6.2E – 4			
Cyanide compounds				
Glycol ethers				
Lead compounds				
Lead	8.0E – 5	1.5E + 0	1.5E + 0	Yes/NA
Manganese compounds				
Manganese		4.0E – 1		
Mercury compounds				
Mercury		3.0E – 1	3.0E + 1	Yes/NA
Fine mineral fibers				
Nickel compounds				
Nickel	2.6E – 4	2.4E – 1	1.0E + 1	
Polycyclic organic matter				

■ **TABLE E-1** (*Continued*)

Substance	Cancer Potency Value Unit Risk	Noncancer Chronic AEL (μg/m³)	Noncancer Acute AEL (μg/m³)	Multi-pathway Exposure/ Oral Potency (mg/kg-d)$^{-1}$
PAH [benzo(*a*) pyrene]	1.7E – 3			Yes/11.5
Radionuclides (including radon)				
Selenium compounds				
Selenium	1.4E – 4	5.0E – 1	2.0E + 0	

[a]Federal HAPs for which the State of California has promulgated risk data have values entered.
[b]NA, not available.
Source: CAPCOA, 1991.

APPENDIX F

LIST OF NSPS SOURCES

■ **TABLE F-1** List of NSPS Sources and Dates for Determining Applicability of Clean Air Act New Source Performance Standards as of August 1, 1995

40 CFR 60 Subpart	Source Category	Proposal Date	Promulgation Date
Ca	Emissions guidelines and compliance times for municipal waste combustors	July 7, 1987	February 11, 1991
Cb	Emission guidelines and compliance times for sulfuric acid production units	July 7, 1987	February 11, 1991
D	Standards of performance for fossil-fuel fired steam generators for which construction is commenced after August 17, 1971	August 17, 1971	December 23, 1971
Da	Standards of performance for electric utility steam generating units for which construction is commenced after September 18, 1978	September 18, 1978	June 11, 1979
Db	Standards of performance for industrial–commercial–institutional steam-generating units	June 19, 1984	November 25, 1986
Dc	Standards of performance for small industrial–commercial–institutional steam-generating units	June 9, 1989	September 12, 1990
E	Standards of performance for incinerators	August 17, 1971	December 23, 1971
Ea	Standards of performance for municipal waste combustors	July 7, 1987	February 11, 1991

■ **TABLE F-1 List of NSPS Sources and Dates for Determining Applicability of Clean Air Act New Source Performance Standards as of August 1, 1995 *(Continued)***

40 CFR 60 Subpart	Source Category	Proposal Date	Promulgation Date
F	Standards of performance for Portland cement plants	August 17, 1971	December 23, 1971
G	Standards of performance for nitric acid plants	August 17, 1971	December 23, 1971
H	Standards of performance for sulfuric acid plants	August 17, 1971	December 23, 1971
I	Standards of performance for asphalt concrete plants	June 11, 1973	March 8, 1974
J	Standards of performance for petroleum refineries	June 11, 1973	March 8, 1974
K	Standards of performance for storage vessels for petroleum liquids for which construction, reconstruction, or modification commenced after June 11, 1973 and prior to May 19, 1978	June 11, 1973	March 8, 1974
Ka	Standard of performance for storage vessels for petroleum liquids for which construction, reconstruction, or modification commenced after May 18, 1978 and prior to July 23, 1984	May 18, 1978	April 4, 1980
Kb	Standards of performance for volatile organic liquid storage vessels	July 23, 1984	April 8, 1987

■ **TABLE F-1** *(Continued)*

40 CFR 60 Subpart	Source Category	Proposal Date	Promulgation Date
	(including petroleum liquid storage vessels) for which construction, reconstruction, or modification commenced after July 23, 1984		
L	Standards of performance for secondary lead smelters	June 11, 1973	March 8, 1974
M	Standards of performance for primary emissions from basic oxygen process furnaces for which construction is commenced after June 11, 1973	June 11, 1973	March 8, 1974
N	Standards of performance for primary emissions from basic oxygen process furnaces for which construction is commenced after June 11, 1973	June 11, 1973	March 8, 1974
Na	Standards of performance for secondary emissions from basic oxygen process steel-making facilities for which construction is commenced after January 20, 1983	January 20, 1983	January 2, 1986
O	Standards of performance for sewage treatment plants	June 11, 1973	March 8, 1974

■ **TABLE F-1** **List of NSPS Sources and Dates for Determining Applicability of Clean Air Act New Source Performance Standards as of August 1, 1995** ***(Continued)***

40 CFR 60 Subpart	Source Category	Proposal Date	Promulgation Date
P	Standards of performance for primary copper smelters	October 16, 1974	January 15, 1976
Q	Standards of performance for primary zinc smelters	October 16, 1974	January 15, 1976
R	Standards of performance for primary lead smelters	October 16, 1974	January 15, 1976
S	Standards of performance for primary aluminum reduction plants	October 23, 1974	January 26, 1976
T	Standards of performance for the phosphate fertilizer industry: wet-process phosphoric acid plants	October 22, 1974	August 6, 1975
U	Standards of performance for the phosphate fertilizer industry: superphosphoric acid plants	October 22, 1974	August 6, 1975
V	Standards of performance for the phosphate fertilizer industry: diammonium phosphate plants	October 22, 1974	August 6, 1975
W	Standards of performance for the phosphate fertilizer industry: triple superphosphate plants	October 22, 1974	August 6, 1975

■ **TABLE F-1** *(Continued)*

40 CFR 60 Subpart	Source Category	Proposal Date	Promulgation Date
X	Standards of performance for the phosphate fertilizer industry: granular triple superphosphate storage facilities	October 22, 1974	August 6, 1975
Y	Standards of performance for coal preparation plants	October 24, 1974	January 15, 1976
Z	Standards of performance for ferroalloy production facilities	October 21, 1974	May 20, 1976
AA	Steel plants: electric arc furnaces	October 21, 1974	September 23, 1975
AAa	Steel plants: electric arc furnaces and argon–oxygen decarburization vessels constructed after August 17, 1983	August 17, 1983	October 31, 1984
BB	Kraft pulp mills	September 24, 1976	February 23, 1978
CC	Glass manufacturing plants	June 15, 1979	October 7, 1980
DD	Grain elevators	January 13, 1977	August 3, 1978
EE	Surface coating of metal furniture	November 28, 1980	October 29, 1982
GG	Stationary gas turbines	October 3, 1977	September 10, 1979
HH	Lime manufacturing plants	May 3, 1977	March 7, 1978
KK	Lead–acid battery manufacturing plants	January 14, 1980	April 16, 1982
LL	Metallic mineral processing plants	August 24, 1982	February 24, 1984
MM	Automobile and light-duty truck surface coating operations	October 5, 1979	December 24, 1980

■ **TABLE F-1 List of NSPS Sources and Dates for Determining Applicability of Clean Air Act New Source Performance Standards as of August 1, 1995 *(Continued)***

40 CFR 60 Subpart	Source Category	Proposal Date	Promulgation Date
NN	Phosphate rock plants	September 21, 1979	April 16, 1982
PP	Ammonium sulfate manufacture plants	April 10, 1980	November 12, 1980
QQ	Graphic arts industry: publication rotogravure printing	October 28, 1980	November 8, 1982
RR	Pressure-sensitive tape and label surface coating operations	December 30, 1980	October 18, 1983
SS	Industrial surface coating: large appliances	December 24, 1980	October 27, 1982
TT	Metal coil surface coating	January 5, 1981	November 1, 1982
UU	Asphalt processing and asphalt roofing manufacture	November 18, 1980; May 26, 1981	August 6, 1982
VV	Equipment leaks of VOC in the synthetic organic chemicals manufacturing industry	January 5, 1981	October 18, 1983
WW	Beverage can surface coating industry	November 26, 1980	August 25, 1983
XX	Bulk gasoline terminals	December 17, 1980	August 18, 1983
AAA	New residential wood heaters	February 18, 1987	February 26, 1988
BBB	Rubber tire manufacturing industry	January 20, 1983	September 15, 1987

■ **TABLE F-1** *(Continued)*

40 CFR 60 Subpart	Source Category	Proposal Date	Promulgation Date
DDD	Polymer manufacturing industry	September 30, 1987	December 11, 1990
FFF	Flexible vinyl and urethane coating and printing	January 18, 1983; October 11, 1983	June 28, 1984
GGG	Equipment leaks of VOC in petroleum refineries	January 4, 1983	May 30, 1984
HHH	Synthetic fiber production facilities	July 13, 1983	April 5, 1984
III	Synthetic organic chemical manufacturing industry air oxidation unit processes	October 21, 1983	June 29, 1990
JJJ	Petroleum dry cleaners	December 14, 1982	September 21, 1984
KKK	Onshore natural gas processing plants	January 24, 1984	June 24, 1985
LLL	Onshore natural gas processing: SO_2 emissions	January 20, 1984	October 1, 1985
NNN	Synthetic organic chemical manufacturing industry distillation operations	December 30, 1983	June 29, 1990
OOO	Nonmetallic mineral processing plants	August 31, 1983	August 1, 1985
PPP	Wool fiberglass insulation manufacturing plants	February 7, 1984	February 25, 1985
QQQ	Petroleum refinery wastewater system VOC emissions	May 4, 1987	November 23, 1988
SSS	Magnetic tape coating facilities	January 22, 1986	October 3, 1988

■ **TABLE F-1 List of NSPS Sources and Dates for Determining Applicability of Clean Air Act New Source Performance Standards as of August 1, 1995** ***(Continued)***

40 CFR 60 Subpart	Source Category	Proposal Date	Promulgation Date
TTT	Industrial surface coating: plastic parts for business machines	January 8, 1986	January 29, 1988
VVV	Polymeric coating of supporting substrates facilities	April 30, 1987	September 11, 1989

APPENDIX

G

LIST OF FEDERAL HAZARDOUS AIR POLLUTANTS

■ **TABLE G-1** **Section 112(b) List of Hazardous Air Pollutants as Listed by the 1990 Federal Clean Air Act**

CAS Number	Pollutant Name
75070	Acetaldehyde
60355	Acetamide
75058	Acetonitrile
98862	Acetophenone
53963	2-Acetylaminofluorene
107028	Acrolein
79061	Acrylamide
79107	Acrylic acid
107131	Acrylonitrile
107051	Allyl chloride
92671	4-Aminobiphenyl
62533	Aniline
90040	*o*-Anisidine
1332214	Asbestos
71432	Benzene (including benzene from gasoline)
92875	Benzidine
98077	Benzotrichloride
100447	Benzyl chloride
92524	Biphenyl
117817	Bis(2-ethylhexyl)phthalate (DEHP)
542881	Bis(chloromethyl)ether
75252	Bromoform
106990	1,3-Butadiene
156627	Calcium cyanamide
105602	Caprolactam
133062	Captan
63252	Carbaryl
75150	Carbon disulfide
56235	Carbon tetrachloride
463581	Carbonyl sulfide
120809	Catecol
133904	Chloramben

■ TABLE G-1 Section 112(b) List of Hazardous Air Pollutants as Listed by the 1990 Federal Clean Air Act (*Continued*)

CAS Number	Pollutant Name
57749	Chlordane
7782505	Chlorine
79118	Chloroacetic acid
532274	2-Chloroacetophenone
108907	Chlorobenzene
510156	Chlorobenzilate
67663	Chloroform
107302	Chloromethyl methyl ether
126998	Chloroprene
1319773	Cresols/cresylic acid (isomers and mixture)
95487	*o*-Cresol
108394	*m*-Cresol
106445	*p*-Cresol
98828	Cumene
94757	2,4-D (2,4-dichlorophenoxyacetic acid) salts and esters
72559	DDE
334883	Diazomethane
132649	Dibenzofurans
96128	1,2-Dibromo-3-chloropropane
84742	Dibutylphthalate
106467	1,4-Dichlorobenzene (*p*)
91941	3,3-Dichlorobenzidine
111444	Dichloroethyl ether (bis(2-chloroethyl)ether)
542756	1,3-Dichloropropene
62737	Dichlorvos
111422	Diethanolamine
121697	*N,N*-Diethyl aniline (*N,N*-dimethylaniline)
64675	Diethyl sulfate
119904	3,3-Dimethoxybenzidine
60117	Dimethyl aminoazobenzene
119937	3,3′-Dimethyl benzidine
79447	Dimethyl carbamoyl chloride
68122	Dimethyl formamide
57147	1,1-Dimethyl hydrazine

■ **TABLE G-1** (*Continued*)

CAS Number	Pollutant Name
131113	Dimethyl phthalate
77781	Dimethyl sulfate
534521	4,6-Dinitro-*o*-cresol and salts
51285	2,4-Dinitrophenol
121142	2,4-Dinitrotoluene
123911	1,4-Dioxane (1,4-diethyleneoxide)
122667	1,2-Diphenylhydrazine
106898	Epichlorohydrin (1-chloro-2,3-epoxypropane)
106887	1,2-Epoxybutane
140885	Ethyl acrylate
100414	Ethyl benzene
51796	Ethyl carbamate (urethane)
75003	Ethyl chloride (chloroethane)
106934	Ethylene dibromide (dibromoethane)
107062	Ethylene dichloride (1,2-dichloroethane)
107211	Ethylene glycol
151564	Ethylene imine (aziridine)
75218	Ethylene oxide
96457	Ethylene thiourea
75343	Ethylidene dichloride (1,1-dichloroethane)
50000	Formaldehyde
76448	Heptachlor
118741	Hexachlorobenzene
87683	Hexachlorobutadiene
77474	Hexachlorocyclopentadiene
67721	Hexachloroethane
822060	Hexamethylene-1,6-diisocyanate
680319	Hexamethylphosphoramide
110543	Hexane
302012	Hydrazine
7647010	Hydrochloric acid [hydrogen chloride (gas only)]
7664393	Hydrogen fluoride (hydrofluoric acid)
123319	Hydroquinone

■ TABLE G-1 **Section 112(b) List of Hazardous Air Pollutants as Listed by the 1990 Federal Clean Air Act (*Continued*)**

CAS Number	Pollutant Name
78591	Isophorone
58899	Lindane
10831	Maleic anhydride
67561	Methanol
72435	Methoxychlor
74839	Methyl bromide (bromomethane)
74873	Methyl chloride (chloromethane)
71556	Methyl chloroform (1,1,1-trichloroethane)
78933	Methyl ethyl ketone (2-butanone)
60344	Methyl hydrazine
74884	Methyl iodide (iodomethane)
108101	Methyl isobutyl ketone (hexone)
624839	Methyl isocyanate
80626	Methyl methacrylate
1634044	Methyl tert butyl ether
101144	4,4′-Methylenebis(2-chloroaniline)
75092	Methylene chloride (dichloromethane)
101688	Methylene diphenyl diisocyanate (MDI)
101779	4,4-Methylenedianiline
91203	Naphthalene
98953	Nitrosobenzene
92933	4-Nitrobiphenyl
100027	4-Nitrophenol
79469	2-Nitropropane
684935	*N*-Nitroso-*N*-methylurea
62759	*N*-Nitrosodimethylamine
59892	*N*-Nitrosomorpholine
56382	Parathion
82688	Pentachloronitrobenzene
87865	Pentachlorophenol

■ **TABLE G-1** *(Continued)*

CAS Number	Pollutant Name
108952	Phenol
106503	*p*-Phenylenediamine
75445	Phosgene
7803512	Phosphine
7723140	Phosphorus
85449	Phthalic anhydride
1336363	Polychlorinated biphenyls (Aroclors)
1120714	1,3-Propane sultone
57578	beta-Propiolactone
123386	Propionaldehyde
114261	Propoxur (Baygon)
78875	Propylene dichloride (1,2-dichloropropane)
75569	Propylene oxide
75558	1,2-Propylenimine (2-methyl aziridine)
91225	Quinoline
106514	Quinone
100425	Styrene
96093	Styrene oxide
1746016	2,3,7,8-Tetrachlorodibenzo-*p*-dioxin
79345	1,1,2,2-Tetrachloroethane
127184	Tetrachloroethylene (perchloroethylene)
7550450	Titanium tetrachloride
108883	Toluene
95807	2,4-Toluene diamine
584849	2,4-Toluene diisocyanate
95534	*o*-Toluidine
8001352	Toxaphene (chlorinated camphene)
120821	1,2,4-Trichlorobenzene
79005	1,1,2-Trichloroethane
79016	Trichloroethylene
95954	2,4,5-Trichlorophenol
88062	2,4,6-Trichlorophenol

■ TABLE G-1 Section 112(b) List of Hazardous Air Pollutants as Listed by the 1990 Federal Clean Air Act (*Continued*)

CAS Number	Pollutant Name
121448	Triethylamine
1582098	Trifluralin
540841	2,2,4-Trimethylpentane
108054	Vinyl acetate
593602	Vinyl bromide
75014	Vinyl chloride
75354	Vinylidene chloride (dichloroethylene)
1330207	Xylenes (isomers and mixture)
95476	*o*-Xylenes
108383	*m*-Xylenes
106423	*p*-Xylenes
	Antimony compounds
	Arsenic compounds (inorganic including arsine)
	Beryllium compounds
	Cadmium compounds
	Chromium compounds
	Cobalt compounds
	Coke oven emissions
	Cyanide compounds[a]
	Glycol ethers[b]
	Lead compounds
	Manganese compounds
	Mercury compounds
	Fine mineral fibers[c]
	Nickel compounds
	Polycyclic organic matter[d]
	Radionuclides (including radon)[e]
	Selenium compounds

Note: For all listings above which contain the word "compounds" and for glycol ethers, the following applies: Unless otherwise specified, these listings are defined as including any unique chemical substance that contains the named chemical (i.e., antimony, arsenic, etc.) as part of that chemical's infrastructure.

[a]X′CN, where X = H′ or any other group where a formal dissociation may occur. For example, KCN or $Ca(CN)_2$.

[b]Includes mono- and diethers of ethylene glycol, diethylene glycol, and triethylene glycol $R-(OCH_2CH_2)_n-OR'$, where

$n = 1, 2,$ or 3

R = alkyl or aryl groups

R′ = R, H, or groups which, when removed, yield glycol ethers with structure; $R-(OCH2CH)_n-OH$ polymers are excluded from the glycol category.

[c]Includes mineral fiber emissions from facilities manufacturing or processing glass, rock, or slag fibers (or other mineral derived fibers) of average diameter 1 micrometer or less.

[d]Includes organic compounds with more than one benzene ring, and which have a boiling point greater than or equal to 100°C.

[e]A type of atom which spontaneously undergoes radioactive decay.

■ TABLE G-2 Draft Corrected List of Section 112 Hazardous Air Pollutants as of September 15, 1993

This draft list includes technical corrections and clarifications under USEPA work group consideration as of September 15, 1993. The official list of Section 112 Hazardous Air Pollutants can be found in the 1990 Federal Clean Air Act, which has been reproduced in Table G-1. The technical corrections and clarifications below have no impact on the source category emissions standards currently in progress.

Chemical Abstracts Service Number	Pollutant
75-07-0	Acetaldehyde
60-35-5	Acetamide
75-05-8	Acetonitrile
98-86-2	Acetophenone
53-96-3	2-Acetylaminofluorene
107-02-8	Acrolein
79-06-1	Acrylamide
79-10-7	Acrylic acid
107-13-1	Acrylonitrile
107-05-1	Allyl chloride
92-67-1	4-Aminobiphenyl
62-53-3	Aniline
90-04-0	*o*-Anisidine
1332-21-4	Asbestos
71-43-2	Benzene (including benzene from gasoline)
92-87-5	Benzidine
98-07-7	Benzotrichloride

■ TABLE G-2 Draft Corrected List of Section 112 Hazardous Air Pollutants as of September 15, 1993 (*Continued*)

Chemical Abstracts Service Number	Pollutant
100-44-7	Benzyl chloride
92-52-4	Biphenyl
117-81-7	Bis(2-ethylhexyl)phthalate (DEHP)
542-88-1	Bis(chloromethyl)ether
75-25-2	Bromoform
106-99-0	1,3-Butadiene
156-62-7	Calcium cyanamide
105-60-2	Caprolactam
133-06-2	Captan
63-25-2	Carbaryl
75-15-0	Carbon disulfide
56-23-5	Carbon tetrachloride
463-58-1	Carbonyl sulfide
120-80-9	Catechol
133-90-4	Chloramben
57-74-9	Chlordane
7782-50-5	Chlorine
79-11-8	Chloroacetic acid
532-27-4	2-Chloroacetophenone
108-90-7	Chlorobenzene
510-15-6	Chlorobenzilate
67-66-3	Chloroform
107-30-2	Chloromethyl methyl ether
126-99-8	Chloroprene
1319-77-3	Cresol/cresylic acid (mixed isomers)
95-48-7	*o*-Cresol
108-39-4	*m*-Cresol
106-44-5	*p*-Cresol
98-82-8	Cumene
	2,4-D (2,4-Dichlorophenoxyacetic acid) (including salts and esters)
72-55-9	DDE [1,1-dichloro-2,2-bis(*p*-chlorophenyl)ethylene]
334-88-3	Diazomethane

■ **TABLE G-2** *Continued)*

Chemical Abstracts Service Number	Pollutant
132-64-9	Dibenzofuran
96-12-8	1,2-Dibromo-3-chloropropane
84-74-2	Dibutyl phthalate
106-46-7	1,4-Dichlorobenzene
91-94-1	3,3′-Dichlorobenzidine
111-44-4	Dichloroethyl ether [bis(2-chloroethyl)ether]
542-75-6	1,3-Dichloropropene
62-73-7	Dichlorvos
111-42-2	Diethanolamine
64-67-5	Diethyl sulfate
119-90-4	3,3′-Dimethoxybenzidine
60-11-7	4-Dimethylaminoazobenzene
121-69-7	*N*,*N*-Dimethylaniline
119-93-7	3,3′-Dimethylbenzidine
79-44-7	Dimethylcarbamoyl chloride
68-12-2	*N*,*N*-Dimethylformamide
57-14-7	1,1-Dimethylhydrazine
131-11-3	Dimethyl phthalate
77-78-1	Dimethyl sulfate
	4,6-Dinitro-*o*-cresol (including salts)
51-28-5	2,4-Dinitrophenol
121-14-2	2,4-Dinitrotoluene
123-91-1	1,4-Dioxane (1,4-diethyleneoxide)
122-66-7	1,2-Diphenylhydrazine
106-89-8	Epichlorohydrin (1-chloro-2,3-epoxypropane)
106-88-7	1,2-Epoxybutane
140-88-5	Ethyl acrylate
100-41-4	Ethylbenzene
51-79-6	Ethyl carbamate (urethane)
75-00-3	Ethyl chloride (chloroethane)
106-93-4	Ethylene dibromide (dibromoethane)
107-06-2	Ethylene dichloride (1,2-dichloroethane)
107-21-1	Ethylene glycol
151-56-4	Ethylene imine (aziridine)
75-21-8	Ethylene oxide

■ **TABLE G-2** Draft Corrected List of Section 112 Hazardous Air Pollutants as of September 15, 1993 (*Continued*)

Chemical Abstracts Service Number	Pollutant
96-45-7	Ethylene thiourea
75-34-3	Ethylidene dichloride (1,1-dichloroethane)
50-00-0	Formaldehyde
76-44-8	Heptachlor
118-74-1	Hexachlorobenzene
87-68-3	Hexachlorobutadiene
	1,2,3,4,5,6-Hexachlorocyclyhexane (all stereoisomers, including lindane)
77-47-4	Hexachlorocyclopentadiene
67-72-1	Hexachloroethane
822-06-0	Hexamethylene diisocyanate
680-31-9	Hexamethylphosphoramide
110-54-3	Hexane
302-01-2	Hydrazine
7647-01-0	Hydrochloric acid [hydrogen chloride (gas only)]
7664-39-3	Hydrogen fluoride (hydrofluoric acid)
123-31-9	Hydroquinone
78-59-1	Isophorone
108-31-6	Maleic anhydride
67-56-1	Methanol
72-43-5	Methoxychlor
74-83-9	Methyl bromide (bromomethane)
74-87-3	Methyl chloride (chloromethane)
71-55-6	Methyl chloroform (1,1,1-trichloroethane)
78-93-3	Methyl ethyl ketone (2-butanone)
60-34-4	Methylhydrazine
74-88-4	Methyl iodide (iodomethane)
108-10-1	Methyl isobutyl ketone (hexone)
624-83-9	Methyl isocyanate
80-62-6	Methyl methacrylate
1634-04-4	Methyl tert-butyl ether
101-14-4	4,4′-Methylenebis(2-chloroaniline)
75-09-2	Methylene chloride (dichloromethane)
101-68-8	4,4′-Methylenediphenyl diisocyanate (MDI)
101-77-9	4,4′-Methylenedianiline

■ **TABLE G-2** *(Continued)*

Chemical Abstracts Service Number	Pollutant
91-20-3	Naphthalene
98-95-3	Nitrobenzene
92-93-3	4-Nitrobiphenyl
100-02-7	4-Nitrophenol
79-46-9	2-Nitropropane
684-93-5	*N*-Nitroso-*N*-methylurea
62-75-9	*N*-Nitrosodimethylamine
59-89-2	*N*-Nitrosomorpholine
56-38-2	Parathion
82-68-8	Pentachloronitrobenzene (quintobenzene)
87-86-5	Pentachlorophenol
108-95-2	Phenol
106-50-3	*p*-Phenylenediamine
75-44-5	Phosgene
7803-51-2	Phosphine
7723-14-0	Phosphorus
85-44-9	Phthalic anhydride
1336-36-3	Polychlorinated biphenyls (Aroclors)
1120-71-4	1,3-Propane sultone
57-57-8	beta-Propiolactone
123-38-6	Propionaldehyde
114-26-1	Propoxur (Baygon)
78-87-5	Propylene dichloride (1,2-dichloropropane)
75-56-9	Propylene oxide
75-55-8	1,2-Propylenimine (2-methylaziridine)
91-22-5	Quinoline
106-51-4	Quinone (*p*-benzoquinone)
100-42-5	Styrene
96-09-3	Styrene oxide
1746-01-6	2,3,7,8-Tetrachlorodibenzo-*p*-dioxin
79-34-5	1,1,2,2-Tetrachloroethane
127-18-4	Tetrachloroethylene (perchloroethylene)
7550-45-0	Titanium tetrachloride
108-88-3	Toluene
95-80-7	Toluene-2,4-diamine
584-84-9	2,4-Toluene diisocyanate

TABLE G-2 Draft Corrected List of Section 112 Hazardous Air Pollutants as of September 15, 1993 (*Continued*)

Chemical Abstracts Service Number	Pollutant
95-53-4	*o*-Toluidine
8001-35-2	Toxaphene (chlorinated camphene)
120-82-1	1,2,4-Trichlorobenzene
79-00-5	1,1,2-Trichloroethane
79-01-6	Trichloroethylene
95-95-4	2,4,5-Trichlorophenol
88-06-2	2,4,6-Trichlorophenol
121-44-8	Triethylamine
1582-09-8	Trifluralin
540-84-1	2,2,4-Trimethylpentane
108-05-4	Vinyl acetate
593-60-2	Vinyl bromide
75-01-4	Vinyl chloride
75-35-4	Vinylidene chloride (1,1-dichloroethylene)
1330-20-7	Xylene (mixed isomers)
95-47-6	*o*-Xylene
108-38-3	*m*-Xylene
106-42-3	*p*-Xylene
	Antimony compounds
	Arsenic compounds (inorganic including arsine)
	Beryllium compounds
	Cadmium compounds
	Chromium compounds
	Cobalt compounds
	Coke oven emissions
	Cyanide compounds[a]
	Glycol ethers[b]
	Lead compounds
	Manganese compounds
	Mercury compounds
	Fine mineral fibers[c]
	Nickel compounds
	Polycyclic organic matter[d]
	Radionuclides (including radon)[e]
	Selenium compounds

Note: For all listings above which contain the word "compounds" and for glycol ethers, the following applies: Unless otherwise specified, these listings are defined as including any unique chemical substance that contains the named chemical (i.e., antimony, arsenic, etc.) as part of that chemical's infrastructure.

[a]X′CN, where X = H′ or any other group where a formal dissociation may occur. For example, KCN or $Ca(CN)_2$.

[b]$R\text{-}(OCH_2CH_2)_n\text{-}OR'$, where:

n = 1, 2, or 3
R = alkyl C7 or less
or R = phenyl or alkyl substituted phenyl
R′ = H, or alkyl C7 or less
or ester
sulfate
phosphate
nitrate
sulfonate

[c]Includes mineral fiber emissions from facilities manufacturing or processing glass, rock, or slag fibers (or other mineral derived fibers) of average diameter 1 micrometer or less.

[d]Includes substituted and/or unsubstituted polycyclic aromatic hydrocarbons and aromatic heterocyclic compounds, with two or more fused rings, at least one of which is benzenoid (i.e., containing six carbon atoms and is aromatic) in structure. Polycyclic organic matter is a mixture of organic compounds containing one or more of these polycyclic aromatic chemicals. Polycyclic organic matter is generally formed or emitted during thermal processes including (1) incomplete combustion, (2) pyrolysis, (3) the volatilization, distillation, or processing of fossil fuels or bitumens, or (4) the distillation or thermal processing of nonfossil fuels. The Administrator may delineate, by test method, what is included in polycyclic organic matter.

[e]A type of atom which spontaneously undergoes radioactive decay.

APPENDIX H

LIST OF SOURCE CATEGORIES EMITTING HAZARDOUS AIR POLLUTANTS

TABLE H-1 Categories of Sources of Hazardous Air Pollutants and Regulation Promulgation Schedule by Industry Group

Category	Schedule Date
Industry Group	
Source category[a]	
Fuel Combustion	
Engine test facilities	11/15/00
Industrial boilers[b]	11/15/00
Institutional/commercial boilers[b]	11/15/00
Process heaters	11/15/00
Stationary internal combustion engines[b]	11/15/97
Stationary turbines[b]	11/15/97
Nonferrous Metals Processing	
Primary aluminum production	11/15/97
Secondary aluminum production	11/15/97
Primary copper smelting	11/15/97
Primary lead smelting	11/15/97
Secondary lead smelting	11/15/94
Lead acid battery manufacturing	11/15/00
Primary magnesium refining	11/15/00
Ferrous Metals Processing	
Coke by-product plants	11/15/00
Coke ovens: charging, top side, and door leaks	12/31/92
Coke ovens: pushing, quenching, and battery stacks	11/15/00
Ferroalloys production	11/15/97
Integrated iron and steel manufacturing	11/15/97
Non-stainless steel manufacturing—electric arc furnace (EAF) operation	11/15/97
Stainless steel manufacturing—electric arc furnace (EAF) operation	11/15/97
Iron foundries	11/15/97
Steel foundries	11/15/97
Steel pickling—HCl process	11/15/97
Mineral Products Processing	
Alumina processing	11/15/00
Asphalt/coal tar application—metal pipes	11/15/00

■ **TABLE H-1** **Categories of Sources of Hazardous Air Pollutants and Regulation Promulgation Schedule by Industry Group** ***(Continued)***

Category	Schedule Date
Asphalt concrete manufacturing	11/15/00
Asphalt processing	11/15/00
Asphalt roofing manufacturing	11/15/00
Chromium refractories production	11/15/97
Clay products manufacturing	11/15/00
Lime manufacturing	11/15/00
Mineral wool production	11/15/97
Portland cement manufacturing	11/15/97
Taconite iron ore processing	11/15/00
Wool fiberglass manufacturing	11/15/97
Petroleum and Natural Gas Production and Refining	
Oil and natural gas production	11/15/97
Petroleum refineries—catalytic cracking (fluid and other) units, catalytic reforming units, and sulfur plant units	11/15/97
Petroleum refineries—other sources not distinctly listed	11/15/94
Liquids Distribution	
Gasoline distribution (Stage 1)	11/15/94
Organic liquids distribution (nongasoline)	11/15/00
Surface Coating Processes	
Aerospace industries	11/15/94
Auto and light duty truck (surface coating)	11/15/97
Flat wood paneling (surface coating)	11/15/00
Large appliance (surface coating)	11/15/00
Magnetic tapes (surface coating)	11/15/94
Manufacture of paints, coatings, and adhesives	11/15/00
Metal can (surface coating)	11/15/00
Metal coil (surface coating)	11/15/00
Metal furniture (surface coating)	11/15/00
Miscellaneous metal parts and products (surface coating)	11/15/00
Paper and other webs (surface coating)	11/15/97
Plastic parts and products (surface coating)	11/15/00
Printing, coating, and dyeing of fabrics	11/15/00
Printing/publishing (surface coating)	11/15/94

■ **TABLE H-1** *(Continued)*

Category	Schedule Date
Shipbuilding and ship repair (surface coating)	11/15/94
Wood furniture (surface coating)	11/15/94
Waste Treatment and Disposal	
Hazardous waste incineration	11/15/00
Municipal landfills	11/15/97
Publicly owned treatment works (POTW) emissions	11/15/95
Sewage sludge incineration	11/15/97
Site remediation	11/15/00
Solid waste treatment, storage and disposal facilities (TSDF)	11/15/94
Agricultural Chemicals Production	
2,4-D salts and esters production	11/15/00
4-Chloro-2-methylphenoxyacetic acid production	11/15/00
4,6-Dinitro-*o*-cresol production	11/15/00
Captafol production[d]	11/15/00
Captan production[d]	11/15/00
Chloroneb production	11/15/97
Chlorothalonil production[d]	11/15/00
Dacthal™ production[d]	11/15/00
Sodium pentachlorophenate production	11/15/00
Tordon™ acid production[d]	11/15/00
Fibers Production Processes	
Acrylic fibers/modacrylic fibers production	11/15/97
Rayon production	11/15/97
Spandex production	11/15/00
Food and Agriculture Processes	
Baker's yeast manufacturing	11/15/00
Cellulose food casing manufacturing	11/15/00
Vegetable oil production	11/15/00
Pharmaceutical Production Processes	
Pharmaceuticals production[d]	11/15/97
Polymers and Resins Production	
Acetal resins production	11/15/97
Acrylonitrile–butadiene–styrene production	11/15/94

■ TABLE H-1 Categories of Sources of Hazardous Air Pollutants and Regulation Promulgation Schedule by Industry Group *(Continued)*

Category	Schedule Date
Alkyd resins production	11/15/00
Amino resins production	11/15/97
Boat manufacturing	11/15/00
Butadiene–furfural cotrimer (R-11)[d]	11/15/00
Butyl rubber production	11/15/94
Carboxymethylcellulose production	11/15/97
Cellophane production	11/15/97
Cellulose ethers production	11/15/00
Epichlorohydrin elastomers production	11/15/94
Epoxy resins production	11/15/94
Ethylene–propylene rubber production	11/15/94
Flexible polyurethane foam production	11/15/97
Hypalon™ production[d]	11/15/94
Maleic anhydride copolymers production	11/15/00
Methylcellulose production	11/15/00
Methyl methacrylate–acrylonitrile–butadiene–styrene production[d]	11/15/94
Methyl methacrylate–butadiene–styrene terpolymers production[d]	11/15/94
Neoprene production	11/15/94
Nitrile butadiene rubber production	11/15/94
Non-nylon polyamides production	11/15/94
Nylon 6 production	11/15/97
Phenolic resins production	11/15/97
Polybutadiene rubber production[d]	11/15/94
Polycarbonates production[d]	11/15/97
Polyester resins production	11/15/97
Polyethylene terephthalate production	11/15/94
Polymerized vinylidene chloride production	11/15/00
Polymethyl methacrylate resins production	11/15/97
Polystyrene production	11/15/94
Polysulfide rubber production[d]	11/15/94
Polyvinyl acetate emulsions production	11/15/97
Polyvinyl alcohol production	11/15/97

TABLE H-1 *(Continued)*

Category	Schedule Date
Polyvinyl butyral production	11/15/97
Polyvinyl chloride and copolymers production	11/15/00
Reinforced plastic composites production	11/15/97
Styrene–acrylonitrile production	11/15/94
Styrene–butadiene rubber and latex production[d]	11/15/94
Production of Inorganic Chemicals	
Ammonium sulfate production—caprolactam by-product plants	11/15/00
Antimony oxides manufacturing	11/15/00
Chlorine production[d]	11/15/97
Chromium chemicals manufacturing	11/15/97
Cyanuric chloride production	11/15/97
Fume silica production	11/15/00
Hydrochloric acid production	11/15/97
Hydrogen cyanide production	11/15/97
Hydrogen fluoride production	11/15/97
Phosphate fertilizers production	11/15/97
Phosphoric acid manufacturing	11/15/97
Quaternary ammonium compounds production	11/15/00
Sodium cyanide production	11/15/97
Uranium hexafluoride production	11/15/00
Production of Organic Chemicals	
Synthetic organic chemical manufacturing	11/15/92
Miscellaneous Processes	
Aerosol can-filling facilities	11/15/97
Benzyltrimethylammonium chloride production	11/15/97
Butadiene dimers production	11/15/97
Carbonyl sulfide production	11/15/00
Chelating agents production	11/15/97
Chlorinated paraffins production[d]	11/15/00
Chromic acid anodizing	11/15/94
Commercial dry cleaning (perchloroethylene)—transfer machines	11/15/92
Commercial sterilization facilities	11/15/94

■ TABLE H-1 Categories of Sources of Hazardous Air Pollutants and Regulation Promulgation Schedule by Industry Group *(Continued)*

Category	Schedule Date
Decorative chromium electroplating	11/15/94
Dodecanedioic acid production[d]	11/15/00
Dry cleaners (petroleum solvent)	11/15/00
Ethylidene norbornene production[d]	11/15/00
Explosives production	11/15/00
Halogenated solvent cleaners	11/15/94
Hard chromium electroplating	11/15/94
Hydrazine production	11/15/97
Industrial cleaning (perchloroethylene)—Dry-to-dry machines	11/15/92
Industrial dry cleaning (perchloroethylene)—transfer machines	11/15/92
Industrial process cooling towers	11/15/94
OBPA/1,3-diisocyanate production[d]	11/15/00
Paint stripper users	11/15/00
Photographic chemicals production	11/15/97
Phthalate plasticizers production	11/15/00
Plywood/particle board manufacturing	11/15/00
Polyether polyols production	11/15/97
Pulp and paper production	11/15/97
Rocket engine test firing	11/15/00
Rubber chemicals manufacturing	11/15/97
Semiconductor manufacturing	11/15/97
Symmetrical tetrachloropyridine production[d]	11/15/00
Tire production	11/15/00
Wood treatment	11/15/97
Categories of Area Sources[c]	
Asbestos processing	11/15/94
Chromic acid anodizing	11/15/94
Commercial dry cleaning (perchloroethylene)—transfer machines	11/15/92
Commercial dry cleaning (perchloroethylene)—dry-to-dry machines	11/15/92
Commercial sterilization facilities	11/15/94

■ **TABLE H-1** *(Continued)*

Category	Schedule Date
Decorative chromium electroplating	11/15/94
Halogenated solvent cleaners	11/15/94
Hard chromium electroplating	11/15/94

[a]Only major sources within any category shall be subject to emission standards under Section 112 unless a finding is made of a threat of adverse effects to human health or the environment for the area sources in a category. All listed categories are exclusive of any specific operations or processes included under other categories that are listed separately.

[b]Sources defined as electric utility steam generating units under Section 112(a)(8) shall not be subject to emission standards pending the findings of the study required under Section 112(n)(1).

[c]A finding of threat of adverse effects to human health or the environment was made for each category of area sources listed above.

[d]The HON, which is scheduled for promulgation by November 15, 1992, includes a negotiated standard for equipment leaks from the SOCMI category and 20 non-SOCMI categories (or subsets of these categories). The notice of agreement on negotiated regulation for equipment leaks (56 FR 9315; March 6, 1991) would apply to equipment handling specific chemicals for these categories or subsets of these categories. The specific processes affected within the categories are listed in Section XX.X0(c) of the March 6, 1991 notice.

APPENDIX

I

SCHEDULE OF MACT STANDARDS PROMULGATION

■ **TABLE I-1** EPA Final Timetable for Regulating Categories of Sources of Hazardous Air Pollutants (58 FR 63941; December 3, 1993)

Source Categories with Emission Standards Due by Nov. 15, 1992

Synthetic organic chemical manufacturing (issued February 1994)
Commercial dry-cleaning (perchloroethylene)—transfer machines[a]
Commercial dry-cleaning (perchloroethylene)—dry-to-dry machines[a]
Commercial dry-cleaning (perchloroethylene)—transfer machines (issued September 1993)
Industrial dry-cleaning (perchloroethylene)—transfer machines (issued September 1993)
Industrial dry-cleaning (perchloroethylene)—dry-to-dry machines (issued September 1993)

Source Categories with Emission Standards Due by Nov. 15, 1994

Acrylonitrile–butadiene–styrene production
Aerospace industries
Asbestos processing[a]
Butyl rubber production
Chromic acid anodizing (issued January 1995)
Chromic acid anodizing[a]
Coke ovens: charging, topside, and door leaks (issued October 1993)
Commercial sterilization facilities (issued December 1994)
Commercial sterilization facilities[a]
Decorative chromium electroplating (issued January 1995)
Decorative chromium electroplating[a]
Epichlorohydrin elastomers production
Epoxy resins production (issued March 1995)
Ethylene–propylene rubber production
Gasoline distribution—stage I (issued December 1994)
Halogenated solvent cleaners (issued December 1994)
Halogenated solvent cleaners[a]
Hard chromium electroplating (issued January 1995)
Hard chromium electroplating[a]
Hypalon (TM) production
Industrial process cooling towers (issued September 1994)
Magnetic tapes (issued December 1994)
Methyl methacrylate–acrylonitrile–butadiene–styrene production
Methyl methacrylate–butadiene–styrene terpolymers production

■ TABLE I-1 EPA Final Timetable for Regulating Categories of Sources of Hazardous Air Pollutants (58 FR 63941; December 3, 1993) *(Continued)*

Neoprene production
Nitrile butadiene rubber production
Non-nylon polyamides production (issued March 1995)
Petroleum refineries—other sources not distinctly listed (issued August 1995)
Polyethylene terepthalate production
Polybutadiene rubber production
Polystyrene production
Polysulfide rubber production
Printing/publishing (surface coating)
Secondary lead smelting
Shipbuilding and ship repair (surface coatings)
Solid waste treatment, storage, and disposal facilities (TSDF)
Styrene–acrylonitrile production
Styrene–butadiene rubber and latex production
Tordon (TM) acid production
Wood furniture (surface coating)

Source Categories with Emission Standards Due by Nov. 15, 1997

2,4-D salts and esters production
4,6-Dinitro-*o*-cresol production
4-Chloro-2-methylphenoxyacetic acid production
Acetal resins production
Acrylic fibers/modacrylic fibers production
Amino resins production
Butadiene dimers production
Captafol production
Captan production
Chloroneb production
Chlorine production
Chlorothalonil production
Chromium chemicals manufacturing
Cyanuric chloride production
Dacthal (TM) production
Ferroalloys production

TABLE I-1 *(Continued)*

Flexible polyurethane foam production
Hydrogen cyanide production
Mineral wool production
Non-stainless steel manufacturing—electric ARC furnace (EAF) operation
Nylon 6 production
Oil and natural gas production
Petroleum refineries—catalytic cracking (fluid and other) units, catalytic reforming units, and sulfur plant units
Pharmaceuticals production
Phenolic resins production
Polycarbonates production
Polyether polyols production
Portland cement manufacturing
Primary aluminum production
Primary copper smelting
Primary lead smelting
Publicly owned treatment works (POTW) emissions (CAA mandated promulgation by November 15, 1995)
Pulp and paper production
Reinforced plastic composites production
Secondary aluminum production
Sodium cyanide production
Sodium pentachlorophenate production
Stainless steel manufacturing—electric ARC furnace (EAF) operation
Steel pickling—HCl process
Wood treatment
Wool fiberglass manufacturing

Source Categories with Emission Standards Due by Nov. 15, 2000

Aerosol can-filling facilities
Alkyd resins production
Alumina processing
Ammonium sulfate production—caprolactam by-product plants
Antimony oxides manufacturing
Asphalt concrete manufacturing
Asphalt processing
Asphalt roofing manufacturing

■ TABLE I-1 EPA Final Timetable for Regulating Categories of Sources of Hazardous Air Pollutants (58 FR 63941; December 3, 1993) *(Continued)*

Asphalt/coal tar application—metal pipes
Auto and light duty truck (surface coating)
Baker's yeast manufacturing
Benzyltrimethylammonium chloride production
Boat manufacturing
Butadiene–furfural cotrimer (R-11)
Carbonyl sulfide production
Carboxymethylcellulose production
Cellophane production
Cellulose ethers production
Cellulose food casing manufacturing
Chelating agents production
Chlorinated paraffins production
Chromium refractories production
Clay products manufacturing
Coke by-product plants
Coke ovens: pushing, quenching, and battery stacks
Dodecanedioic acid production
Dry cleaning (petroleum solvent)
Engine test facilities
Ethylidene norbornene production
Explosive production
Flat wood paneling (surface coating)
Fume silica production
Hazardous waste incineration
Hydrazine production
Hydrochloric acid production
Hydrogen fluoride production
Industrial boilers
Institutional/commercial boilers
Integrated iron and steel manufacturing
Iron foundries
Large appliance (surface coating)
Lead acid battery manufacturing

■ **TABLE I-1** *(Continued)*

Lime manufacturing
Maleic anhydride copolymers production
Manufacturing of paints, coatings, and adhesives
Metal can (surface coating)
Metal coil (surface coating)
Metal furniture (surface coating)
Methylcellulose production
Miscellaneous metal parts and products (surface coating)
Municipal landfills
OBPA/1,3-diisocyanate production
Organic-liquids distribution (nongasoline)
Paint stripper users
Paper and other webs (surface coating)
Phosphate fertilizers production
Phosphoric acid manufacturing
Photographic chemicals production
Phthalate plasticizers production
Plastic parts and products (surface coating)
Plywood/particle board manufacturing
Polyester resins production
Polymerized vinylidene chloride production
Polymethyl methacrylate resins production
Polyvinyl acetate emulsions production
Polyvinyl alcohol production
Polyvinyl butyral production
Polyvinyl chloride and copolymers production
Primary magnesium refining
Printing, coating, and dyeing of process fabrics
Process heaters
Quaternary ammonium compounds production
Rayon production
Rocket engine test firing
Rubber chemicals manufacturing
Semiconductor manufacturing
Sewage sludge incineration
Site remediation

■ TABLE I-1 EPA Final Timetable for Regulating Categories of Sources of Hazardous Air Pollutants (58 FR 63941; December 3, 1993) *(Continued)*

Spandex production
Stationary internal combustion engines
Stationary turbines
Steel foundries
Symmetrical tetrachloropyridine production
Taconite iron ore processing
Tire production
Tordon (TM) acid production
Uranium hexafluoride production
Vegetable oil production

[a]Denotes area source category. Some source categories are regulated under both major and area source provisions.

APPENDIX

J

LIST OF CHEMICALS AND THRESHOLD QUANTITIES FOR ACCIDENTAL RELEASE PREVENTION

TABLE J-1 List of Regulated Toxic Substances and Threshold Quantities for Accidental Release Prevention (40 CFR 68.130) [Alphabetical Order—100 Substances]

Chemical Name	CAS No.	Threshold Quantity (Pounds)	Basis for Listing
Acetone cyanohydrin	75-86-5	5,000	*a*
Acrolein	107-02-8	1,000	*a*
Acrylonitrile	107-13-1	10,000	*a*
Acrylyl chloride	814-68-6	1,000	*a*
Allyl alcohol	107-18-6	5,000	*a*
Allylamine	107-11-9	1,000	*a*
Ammonia (anhydrous)	7664-41-7	1,000	*a, b*
Ammonia (aqueous solution, conc. 20% or greater)	7664-41-7	5,000	*a, b*
Aniline	62-53-3	5,000	*a*
Antimony pentafluoride	7783-70-2	1,000	*a*
Arsenous trichloride	7784-34-1	5,000	*a*
Arsine	7784-42-1	500	*a*
Benzal chloride	98-87-3	1,000	*a*
Benzenamine, 3-(trifluoromethyl)-	98-16-8	1,000	*a*
Benzotrichloride	98-07-7	500	*a*
Benzyl chloride	100-44-7	1,000	*a*
Benzyl cyanide	140-29-4	1,000	*a*
Boron trichloride	10294-34-5	1,000	*a*
Boron trifluoride	7637-07-2	1,000	*a*
Boron trifluoride compound with methyl ether (1:1)	353-42-4	5,000	*a*
Bromine	7726-95-6	1,000	*a, b*
Carbon disulfide	75-15-0	10,000	*a*
Chlorine	7782-50-5	1,000	*a, b*
Chlorine dioxide	10049-04-4	500	*c*
Chloroethanol	107-07-3	1,000	*a*
Chloroform	67-66-3	10,000	*a*
Chloromethyl ether	542-88-1	500	*a*
Chloromethyl methyl ether	107-30-2	1,000	*a*
Crotonaldehyde	4170-30-3	10,000	*a*
Crotonaldehyde, (*E*)-	123-73-9	10,000	*a*

■ **TABLE J-1** List of Regulated Toxic Substances and Threshold Quantities for Accidental Release Prevention (40 CFR 68.130) (*Continued*)

Chemical Name	CAS No.	Threshold Quantity (Pounds)	Basis for Listing
Cyanogen chloride	506-77-4	1,000	*c*
Cyclohexylamine	108-91-8	5,000	*a*
Diborane	19287-45-7	500	*a*
Trans-1,4-dichlorobutene	110-57-6	1,000	*a*
Dichloroethyl ether	111-44-4	10,000	*a*
Dimethyldichlorosilane	75-78-5	1,000	*a*
Dimethylhydrazine	57-14-7	5,000	*a*
Dimethyl phosphorochloridothioate	2524-03-0	1,000	*a*
Epichlorohydrin	106-89-8	10,000	*a*
Ethylenediamine	107-15-3	10,000	*a*
Ethylene imine	151-56-4	1,000	*a*
Ethylene oxide	75-21-8	5,000	*a, b*
Fluorine	7782-41-4	500	*a*
Formaldehyde	50-00-0	500	*a*
Formaldehyde cyanohydrin	107-16-4	5,000	*a*
Furan	110-00-9	1,000	*a*
Hydrazine	302-01-2	5,000	*a*
Hydrochloric acid (solution, conc. 25% or greater)	7647-01-0	5,000	*d*
Hydrocyanic acid	74-90-8	500	*a, b*
Hydrogen chloride (anhydrous)	7647-01-0	1,000	*a, b*
Hydrogen fluoride	7664-39-3	500	*a, b*
Hydrogen peroxide (conc. >52%)	7722-84-1	5,000	*a*
Hydrogen selenide	7783-07-5	500	*a*
Hydrogen sulfide	7783-06-4	1,000	*a, b*
Iron, pentacarbonyl-	13463-40-6	500	*a*
Isobutyronitrile	78-82-0	10,000	*a*
Isopropyl chloroformate	108-23-6	5,000	*a*
Lactonitrile	78-97-7	5,000	*a*
Methacrylonitrile	126-98-7	1,000	*a*
Methyl bromide	74-83-9	5,000	*a*

■ **TABLE J-1** (*Continued*)

Chemical Name	CAS No.	Threshold Quantity (Pounds)	Basis for Listing
Methyl chloride	74-87-3	10,000	*b*
Methyl chloroformate	79-22-1	1,000	*a*
Methyl hydrazine	60-34-4	5,000	*a*
Methyl isocyanate	624-83-9	1,000	*a, b*
Methyl mercaptan	74-93-1	1,000	*a*
Methyl thiocyanate	556-64-9	10,000	*a*
Methyltrichlorosilane	75-79-6	1,000	*a*
Nickel carbonyl	13463-39-3	500	*a*
Nitric acid	7697-37-2	5,000	*a*
Nitric oxide	10102-43-9	1,000	*a*
Nitrobenzene	98-95-3	10,000	*e*
Parathion	56-38-2	1,000	*e*
Peracetic acid	79-21-0	1,000	*a*
Perchloromethylmercaptan	594-42-3	1,000	*a*
Phenol (liquid)	108-95-2	10,000	*e*
Phosgene	75-44-5	500	*a, b*
Phosphine	7803-51-2	1,000	*a*
Phosphorus oxychloride	10025-87-3	1,000	*a*
Phosphorus trichloride	7719-12-2	5,000	*a*
Piperidine	110-89-4	5,000	*a*
Propionitrile	107-12-0	1,000	*a*
Propyl chloroformate	109-61-5	5,000	*a*
Propyleneimine	75-55-8	10,000	*a*
Propylene oxide	75-56-9	10,000	*a*
Pyridine, 2-methyl-5-vinyl-	140-76-1	1,000	*a*
Sulfur dioxide	7446-09-5	1,000	*a, b*
Sulfuric acid	7664-93-9	5,000	*e*
Sulfur tetrafluoride	7783-60-0	1,000	*a*
Sulfur trioxide	7446-11-9	1,000	*a, b*
Tetramethyllead	75-74-1	1,000	*a*
Tetranitromethane	509-14-8	1,000	*a*
Thiophenol	108-98-5	1,000	*a*
Titanium tetrachloride	7550-45-0	500	*a*

■ TABLE J-1 List of Regulated Toxic Substances and Threshold Quantities for Accidental Release Prevention (40 CFR 68.130) (*Continued*)

Chemical Name	CAS No.	Threshold Quantity (Pounds)	Basis for Listing
Toluene 2,4-diisocyanate	584-84-9	1,000	*b*
Toluene 2,6-diisocyanate	91-08-7	1,000	*b*
Toluene diisocyanate (unspecified isomer)	26471-62-5	1,000	*b*
Trichloroethylsilane	115-21-9	1,000	*a*
Trimethylchlorosilane	75-77-4	1,000	*a*
Vinyl acetate monomer	108-05-4	5,000	*a*
Vinyl chloride	75-01-4	10,000	*b*

[a]On EHS list, vapor pressure 0.5 mmHg or greater.
[b]Mandated for listing by Congress.
[c]Toxic gas.
[d]Listed based on toxicity of hydrogen chloride, potential to release hydrogen chloride, and history of accidents.
[e]On EHS list, vapor pressure less than 0.5 mmHg, but has been involved in accidents resulting in death or injury.

■ TABLE J-2 List of Regulated Toxic Substances and Threshold Quantities for Accidental Release Prevention (40 CFR 68.130) [CAS Number Order—100 Substances]

CAS No.	Chemical Name	Threshold Quantity (Pounds)	Basis for Listing
50-00-0	Formaldehyde	500	*a*
56-38-2	Parathion	1,000	*b*
57-14-7	Dimethylhydrazine	5,000	*a*
60-34-4	Methyl hydrazine	5,000	*a*
62-53-3	Aniline	5,000	*a*
67-66-3	Chloroform	10,000	*a*
74-83-9	Methyl bromide	5,000	*a*
74-87-3	Methyl chloride	10,000	*c*
74-90-8	Hydrocyanic acid	500	*a, c*
74-93-1	Methyl mercaptan	1,000	*a*
75-01-4	Vinyl chloride	10,000	*c*

■ **TABLE J-2** (*Continued*)

CAS No.	Chemical Name	Threshold Quantity (Pounds)	Basis for Listing
75-15-0	Carbon disulfide	10,000	*a*
75-21-8	Ethylene oxide	5,000	*a, c*
75-44-5	Phosgene	500	*a, c*
75-55-8	Propyleneimine	10,000	*a*
75-56-9	Propylene oxide	10,000	*a*
75-74-1	Tetramethyllead	1,000	*a*
75-77-4	Trimethylchlorosilane	1,000	*a*
75-78-5	Dimethyldichlorosilane	1,000	*a*
75-79-6	Methyltrichlorosilane	1,000	*a*
75-86-5	Acetone cyanohydrin	5,000	*a*
78-82-0	Isobutyronitrile	10,000	*a*
78-97-7	Lactonitrile	5,000	*a*
79-21-0	Peracetic acid	1,000	*a*
79-22-1	Methyl chloroformate	1,000	*a*
91-08-7	Toluene 2,6-diisocyanate	1,000	*c*
98-07-7	Benzotrichloride	500	*a*
98-16-8	Benzenamine, 3-(trifluoromethyl)	1,000	*a*
98-87-3	Benzal chloride	1,000	*a*
98-95-3	Nitrobenzene	10,000	*b*
100-44-7	Benzyl chloride	1,000	*a*
106-89-8	Epichlorohydrin	10,000	*a*
107-02-8	Acrolein	1,000	*a*
107-07-3	Chloroethanol	1,000	*a*
107-11-9	Allylamine	1,000	*a*
107-12-0	Propionitrile	1,000	*a*
107-13-1	Acrylonitrile	10,000	*a*
107-15-3	Ethylenediamine	10,000	*a*
107-16-4	Formaldehyde cyanohydrin	5,000	*a*
107-18-6	Allyl alcohol	5,000	*a*
107-30-2	Chloromethyl methyl ether	1,000	*a*
108-05-4	Vinyl acetate monomer	5,000	*a*
108-23-6	Isopropyl chloroformate	5,000	*a*
108-91-8	Cyclohexylamine	5,000	*a*
108-95-2	Phenol (liquid)	10,000	*b*
108-98-5	Thiophenol	1,000	*a*

■ **TABLE J-2** List of Regulated Toxic Substances and Threshold Quantities for Accidental Release Prevention (40 CFR 68.130) [CAS Number Order—100 Substances] (*Continued*)

CAS No.	Chemical Name	Threshold Quantity (Pounds)	Basis for Listing
109-61-5	Propyl chloroformate	5,000	*a*
110-00-9	Furan	1,000	*a*
110-57-6	*trans*-1,4-Dichlorobutene	1,000	*a*
110-89-4	Piperidine	5,000	*a*
111-44-4	Dichloroethyl ether	10,000	*a*
115-21-9	Trichloroethylsilane	1,000	*a*
123-73-9	Crotonaldehyde, (*E*)-	10,000	*a*
126-98-7	Methacrylonitrile	1,000	*a*
140-29-4	Benzyl cyanide	1,000	*a*
140-76-1	Pyridine, 2-methyl-5-vinyl	1,000	*a*
151-56-4	Ethylene imine	1,000	*a*
302-01-2	Hydrazine	5,000	*a*
353-42-4	Boron trifluoride compound with methyl ether (1:1)	5,000	*a*
506-77-4	Cyanogen chloride	1,000	*d*
509-14-8	Tetranitromethane	1,000	*a*
542-88-1	Chloromethyl ether	500	*a*
556-64-9	Methyl thiocyanate	10,000	*a*
584-84-9	Toluene 2,4-diisocyanate	1,000	*c*
594-42-3	Perchloromethylmercaptan	1,000	*a*
624-83-9	Methyl isocyanate	1,000	*a, c*
814-68-6	Acrylyl chloride	1,000	*a*
2524-03-0	Dimethylphosphorochloridothioate	1,000	*a*
4170-30-3	Crotonaldehyde	10,000	*a*
7446-09-5	Sulfur dioxide	1,000	*a, c*
7446-11-9	Sulfur trioxide	1,000	*a, c*
7550-45-0	Titanium tetrachloride	500	*a*
7637-07-2	Boron trifluoride	1,000	*a*
7647-01-0	Hydrogen chloride (anhydrous)	1,000	*a, c*
7647-01-0	Hydrochloric acid (solution, conc. 25% or greater)	5,000	*e*

TABLE J-2 (*Continued*)

CAS No.	Chemical Name	Threshold Quantity (Pounds)	Basis for Listing
7664-39-3	Hydrogen fluoride	500	*a, c*
7664-41-7	Ammonia (anhydrous)	1,000	*a, c*
7664-41-7	Ammonia (aqueous solution, conc. 20% or greater)	5,000	*a, c*
7664-93-9	Sulfuric acid	5,000	*b*
7697-37-2	Nitric acid	5,000	*a*
7719-12-2	Phosphorus trichloride	5,000	*a*
7722-84-1	Hydrogen peroxide (conc. >52%)	5,000	*a*
7726-95-6	Bromine	1,000	*a, c*
7782-41-4	Fluorine	500	*a*
7782-50-5	Chlorine	1,000	*a, c*
7783-06-4	Hydrogen sulfide	1,000	*a, c*
7783-07-5	Hydrogen selenide	500	*a*
7783-60-0	Sulfur tetrafluoride	1,000	*a*
7783-70-2	Antimony pentafluoride	1,000	*a*
7784-34-1	Arsenous trichloride	5,000	*a*
7784-42-1	Arsine	500	*a*
7803-51-2	Phosphine	1,000	*a*
10025-87-3	Phosphorus oxychloride	1,000	*a*
10049-04-4	Chlorine dioxide	500	*d*
10102-43-9	Nitric oxide	1,000	*a*
10294-34-5	Boron trichloride	1,000	*a*
13463-39-3	Nickel carbonyl	500	*a*
13463-40-6	Iron, pentacarbonyl	500	*a*
19287-45-7	Diborane	500	*a*
26471-62-5	Toluene diisocyanate (unspecified isomer)	1,000	*c*

[a]On EHS list, vapor pressure 0.5 mmHg or greater.

[b]On EHS list, vapor pressure less than 0.5 mmHg, but has been involved in accidents resulting in death or injury.

[c]Mandated for listing by Congress.

[d]Toxic gas.

[e]Listed based on toxicity of hydrogen chloride, potential to release hydrogen chloride, and history of accidents.

■ TABLE J-3 List of Regulated Flammable Substances and Threshold Quantities for Accidental Release Prevention (40 CFR 68.130) [Alphabetical Order—62 Substances]

Chemical Name	CAS No.	Threshold Quantity (Pounds)	Basis for Listing
Acetaldehyde	75-07-0	10,000	*a*
Acetylene	74-86-2	10,000	*b*
Bromotrifluorethylene	598-73-2	10,000	*b*
1,3-Butadiene	106-99-0	10,000	*b*
Butane	106-97-8	10,000	*b*
1-Butene	106-98-9	10,000	*b*
2-Butene	107-01-7	10,000	*b*
Butene	25167-67-3	10,000	*b*
2-Butene-*cis*	590-18-1	10,000	*b*
2-Butene-*trans*	624-64-6	10,000	*b*
Carbon oxysulfide	463-58-1	10,000	*b*
Chlorine monoxide	7791-21-1	10,000	*b*
2-Chloropropylene	557-98-2	10,000	*a*
1-Chloropropylene	590-21-6	10,000	*a*
Cyanogen	460-19-5	10,000	*b*
Cyclopropane	75-19-4	10,000	*b*
Dichlorosilane	4109-96-0	10,000	*b*
Difluoroethane	75-37-6	10,000	*b*
Dimethylamine	124-40-3	10,000	*b*
2,2-Dimethylpropane	463-82-1	10,000	*b*
Ethane	74-84-0	10,000	*b*
Ethyl acetylene	107-00-6	10,000	*b*
Ethylamine	75-04-7	10,000	*b*
Ethyl chloride	75-00-3	10,000	*b*
Ethylene	74-85-1	10,000	*b*
Ethyl ether	60-29-7	10,000	*a*
Ethyl mercaptan	75-08-1	10,000	*a*
Ethyl nitrite	109-95-5	10,000	*b*
Hydrogen	1333-74-0	10,000	*b*
Isobutane	75-28-5	10,000	*b*
Isopentane	78-78-4	10,000	*a*

■ **TABLE J-3** (*Continued*)

Chemical Name	CAS No.	Threshold Quantity (Pounds)	Basis for Listing
Isoprene	78-79-5	10,000	[a]
Isopropylamine	75-31-0	10,000	[a]
Isopropyl chloride	75-29-6	10,000	[a]
Methane	74-82-8	10,000	[b]
Methylamine	74-89-5	10,000	[b]
3-Methyl-1-butene	563-45-1	10,000	[b]
2-Methyl-1-butene	563-46-2	10,000	[a]
Methyl ether	115-10-6	10,000	[b]
Methyl formate	107-31-3	10,000	[a]
2-Methylpropene	115-11-7	10,000	[b]
1,3-Pentadiene	504-60-9	10,000	[b]
Pentane	109-66-0	10,000	[a]
1-Pentene	109-67-1	10,000	[a]
2-Pentene, (*E*)-	646-04-8	10,000	[a]
2-Pentene, (*Z*)-	627-20-3	10,000	[a]
Propadiene	463-49-0	10,000	[b]
Propane	74-98-6	10,000	[b]
Propylene	115-07-1	10,000	[b]
Propyne	74-99-7	10,000	[b]
Silane	7803-62-5	10,000	[b]
Tetrafluoroethylene	116-14-3	10,000	[b]
Tetramethylsilane	75-76-3	10,000	[a]
Trichlorosilane	10025-78-2	10,000	[a]
Trifluoro-chloroethylene	79-38-9	10,000	[b]
Trimethylamine	75-50-3	10,000	[b]
Vinyl acetylene	689-97-4	10,000	[b]
Vinyl ethyl ether	109-92-2	10,000	[a]
Vinyl fluoride	75-02-5	10,000	[b]
Vinylidene chloride	75-35-4	10,000	[a]
Vinylidene fluoride	75-38-7	10,000	[b]
Vinyl methyl ether	107-25-5	10,000	[b]

[a]Volatile flammable liquid.
[b]Flammable gas.

■ **TABLE J-4** List of Regulated Flammable Substances and Threshold Quantities for Accidental Release Prevention (40 CFR 68.130) [CAS Number Order—62 Substances]

CAS No.	Chemical Name	Threshold Quantity (Pounds)	Basis for Listing
60-29-7	Ethyl ether	10,000	a
74-82-8	Methane	10,000	b
74-84-0	Ethane	10,000	b
74-85-1	Ethylene	10,000	b
74-86-2	Acetylene	10,000	b
74-89-5	Methylamine	10,000	b
74-98-6	Propane	10,000	b
74-99-7	Propyne	10,000	b
75-00-3	Ethyl chloride	10,000	b
75-02-5	Vinyl fluoride	10,000	b
75-04-7	Ethylamine	10,000	b
75-07-0	Acetaldehyde	10,000	a
75-08-1	Ethyl mercaptan	10,000	a
75-19-4	Cyclopropane	10,000	b
75-28-5	Isobutane	10,000	b
75-29-6	Isopropyl chloride	10,000	a
75-31-0	Isopropylamine	10,000	a
75-35-4	Vinylidene chloride	10,000	a
75-37-6	Difluoroethane	10,000	b
75-38-7	Vinylidene fluoride	10,000	b
75-50-3	Trimethylamine	10,000	b
75-76-3	Tetramethylsilane	10,000	a
78-78-4	Isopentane	10,000	a
78-79-5	Isoprene	10,000	a
79-38-9	Trifluorochloroethylene	10,000	b
106-97-8	Butane	10,000	b
106-98-9	1-Butene	10,000	b
106-99-0	1,3-Butadiene	10,000	b
107-00-6	Ethyl acetylene	10,000	b
107-01-7	2-Butene	10,000	b
107-25-5	Vinyl methyl ether	10,000	b
107-31-3	Methyl formate	10,000	a

TABLE J-4 (*Continued*)

CAS No.	Chemical Name	Threshold Quantity (Pounds)	Basis for Listing
109-66-0	Pentane	10,000	[a]
109-67-1	1-Pentene	10,000	[a]
109-92-2	Vinyl ethyl ether	10,000	[a]
109-95-5	Ethyl nitrite	10,000	[b]
115-07-1	Propylene	10,000	[b]
115-10-6	Methyl ether	10,000	[b]
115-11-7	2-Methylpropene	10,000	[b]
116-14-3	Tetrafluoroethylene	10,000	[b]
124-40-3	Dimethylamine	10,000	[b]
460-19-5	Cyanogen	10,000	[b]
463-49-0	Propadiene	10,000	[b]
463-58-1	Carbon oxysulfide	10,000	[b]
463-82-1	2,2-Dimethylpropane	10,000	[b]
504-60-9	1,3-Pentadiene	10,000	[b]
557-98-2	2-Chloropropylene	10,000	[a]
563-45-1	3-Methyl-1-butene	10,000	[b]
563-46-2	2-Methyl-1-butene	10,000	[a]
590-18-1	2-Butene-*cis*	10,000	[b]
590-21-6	1-Chloropropylene	10,000	[a]
598-73-2	Bromotrifluorethylene	10,000	[b]
624-64-6	2-Butene-trans	10,000	[b]
627-20-3	2-Pentene, (*Z*)-	10,000	[a]
646-04-8	2-Pentene, (*E*)-	10,000	[a]
689-97-4	Vinyl acetylene	10,000	[b]
1333-74-0	Hydrogen	10,000	[b]
4109-96-0	Dichlorosilane	10,000	[b]
7791-21-1	Chlorine monoxide	10,000	[b]
7803-62-5	Silane	10,000	[b]
10025-78-2	Trichlorosilane	10,000	[a]
25167-67-3	Butene	10,000	[b]

[a]Volatile flammable liquid.

[b]Flammable gas.

APPENDIX

K

TABLE OF SIC MAJOR GROUPS

TABLE K-1 SIC Major Groups

SIC Major Group Code	Primary Function
01	Agricultural production—crops
02	Agricultural production—livestock and animal specialties
07	Agricultural services
08	Forestry
09	Fishing, hunting, and trapping
10	Metal mining
12	Coal mining
13	Oil and gas extraction
14	Mining and quarrying of nonmetallic minerals, except fuels
15	Building construction—general contractors and operative builders
16	Heavy construction other than building construction—contractors
17	Construction—special trade contractors
20	Food and kindred products
21	Tobacco products
22	Textile mill products
23	Apparel and other finished products made from fabrics and similar materials
24	Lumber and wood products, except furniture
25	Furniture and fixtures
27	Printing, publishing, and allied industries
28	Chemicals and allied products
29	Petroleum refining and related industries
30	Rubber and miscellaneous plastics products
31	Leather and leather products
32	Stone, clay, glass, and concrete products
33	Primary metal industries
34	Fabricated metal products, except machinery and transportation equipment
35	Industrial and commercial machinery and computer equipment
36	Electronic and other electrical equipment and components, except computer equipment

■ **TABLE K-1** SIC Major Groups (*Continued*)

SIC Major Group Code	Primary Function
37	Transportation equipment
38	Measuring, analyzing, and controlling instruments; photographic, medical, and optical goods; watches and clocks
39	Miscellaneous manufacturing industries
40	Railroad transportation
41	Local and suburban transit and interurban highway passenger transportation
42	Motor freight transportation and warehousing
43	United States Postal Service
44	Water transportation
45	Transportation by air
46	Pipelines, except natural gas
47	Transportation services
48	Communications
49	Electric, gas, and sanitary services
50	Wholesale trade—durable goods
51	Wholesale trade—nondurable goods
52	Building material, hardware, garden supply, and mobile home dealers
53	General merchandise stores
54	Food stores
55	Automotive dealers and gasoline service stations
56	Apparel and accessory stores
57	Home furniture, furnishings, and equipment stores
58	Eating and drinking places
59	Miscellaneous retail
60	Depository institutions
61	Nondepository credit institutions
62	Security and commodity brokers, dealers, exchanges, and services
63	Insurance carriers
64	Insurance agents, brokers, and service
65	Real estate

■ **TABLE K-1** *(Continued)*

SIC Major Group Code	Primary Function
67	Holding and other investment offices
70	Hotels, rooming houses, camps, and other lodging places
72	Personal services
73	Business services
75	Automotive repair, services, and parking
76	Miscellaneous repair services
78	Motion pictures
79	Amusement and recreation services
80	Health services
81	Legal services
82	Educational services
83	Social services
84	Museums, art galleries, and botanical and zoological gardens
86	Membership organizations
87	Engineering, accounting, research, management, and related services
88	Private households
89	Miscellaneous services
91	Executive, legislative, and general government, except finance
92	Justice, public order, and safety
93	Public finance, taxation, and monetary policy
94	Administration of human resource programs
95	Administration of environmental quality and housing programs
96	Administration of economic programs
97	National security and international affairs
99	Nonclassifiable establishments

APPENDIX

L

NONATTAINMENT AREAS

The following tables list the nonattainment areas and their classifications as of July 9, 1994. The tables are as follows:

- L-1 Ozone Nonattainment Areas by Severity Classification
- L-2 Carbon Monoxide Nonattainment Areas by Severity Classification
- L-3 PM_{10} Nonattainment Areas by State and Severity Classification
- L-4 Sulfur Dioxide Nonattainment Areas by State
- L-5 Nitrogen Dioxide Nonattainment Areas by State
- L-6 Lead Nonattainment Areas by State

Note that within the counties and AQCRs listed, the precise boundaries of the nonattainment areas may vary and the 40 CFR 81.300 *et seq.* regulations must be consulted in order to verify the nonattainment status of an area. Note also that the status of areas regularly changes due to redesignations. Therefore, the most current regulations must be consulted, and discussions with the local permitting authority are advised.

■ TABLE L-1 Ozone Nonattainment Areas by Severity Classification (40 CFR 81.300 *et seq.*, as of July 9, 1994)

Extreme	
Los Angeles–South Coast Air Basin, CA	
Severe—17	
Chicago–Gary–Lake County, IL–IN	Riverside Co, CA
Houston–Galveston–Brazona, TX	San Bernardino Co, CA
Milwaukee–Racine, WI	Southeast Desert Modified–AQMA, CA
New York–N. New Jersey–Long Island, NY–NJ–CT	
Severe—15	
Baltimore, MD	Ventura Co, CA
Philadelphia–Wilm–Trent, PA–NJ–DE–MD	
Serious	
Atlanta, GA	Portsmouth–Dover–Rochester, NH
Baton Rouge, LA	Providence (All RI), RI
Beaumont–Port Arthur, TX	Sacramento Metro, CA

■ TABLE L-1 Ozone Nonattainment Areas by Severity Classification (40 CFR 81.300 *et seq.*, as of July 9, 1994) *(Continued)*

Boston–Lawrence–Worcester (E. MA), MA–NH	San Diego, CA
El Paso, TX	San Joaquin Valley, CA
Greater Connecticut	Solano Co, CA
Placer Co, CA	Springfield (Western MA), MA
	Washington, DC–MD–VA
Moderate	
Atlantic City, NJ	Monterey Bay Area, CA
Charleston, WV	Muskegon, MI
Charlotte–Gastonia, NC	Nashville, TN
Cincinnati–Hamilton, OH–KY	Oldham Co, KY
Cleveland–Akron–Lorain, OH	Parkersburg, WV
Dallas–Fort Worth, TX	Phoenix, AZ
Dayton–Springfield, OH	Pittsburgh–Beaver Valley, PA
Detroit–Ann Arbor, MI	Portland, ME
Grand Rapids, MI	Raleigh–Durham, NC
Greensboro–Winston Salem–H Point, NC	Reading, PA
Huntington–Ashland, WV–KY	Richmond–Petersburg, VA
Jefferson Co, KY	Saginaw–Bay City Area, MI
Kewaunee Co, WI	Salt Lake City, UT
Knox and Lincoln Cos, ME	San Francisco–Bay Area, CA
Lewiston–Auburn, ME	Santa Barbara–Santa Maria–Lompoc, CA
Louisville, KY–IN	St. Louis, MO–IL
Manitowoc Co, WI	Sheboygan Co, WI
Miami–Fort Lauderdale–W. Palm Beach, FL	Toledo, OH
Marginal	
Albany–Schenectady–Troy, NY	Lancaster, PA
Allentown–Bethlehem–Easton, PA–NJ	Lexington–Fayette, KY
Altoona, PA	Manchester, NH
Birmingham, AL	Memphis, TN
Buffalo–Niagara Falls, NY	Norfolk–Virginia Beach–Newport News, VA

■ **TABLE L-1** *(Continued)*

Canton, OH
Cherokee Co, SC
Columbus, OH
Door Co, WI
Edmonson Co, KY
Erie, PA
Essex Co (Whiteface Mtn), NY
Evansville, IN
Greenbrier Co, WV
Hancock and Walso Cos, ME
Hamburg–Lebanon–Carlisle, PA
Indianapolis, IN
Jefferson Co, NY
Jersey Co, IL
Johnstown, PA
Kent and Queen Anne's Cos, MD
Knoxville, TN
Lake Charles, LA
Owensboro, KY
Paducah, KY
Portland–Vancouver AQMA, OR–WA
Poughkeepsie, NY
Reno, NV
Scranton–Wilkes-Barre, PA
Seattle–Tacoma, WA
Smyth Co, VA (White Top Mtn)
South Bend–Elkhart, IN
Sussex Co, DE
Tampa–St. Petersburg–Clearwater, FL
Walworth Co, WI
York, PA
Youngstown–Warren–Sharon, OH–PA

Transitional

Chico, CA
Clinton Co, OH
Denver–Boulder, CO
Duval Co, FL
Flint, MI
Imperial Co, CA
Lafayette, LA
Lansing, MI
New Orleans, LA
Preble Co, OH
Steubenville Co, OH
Sutter Co, CA
Yuba Co, CA

Incomplete Data

Beauregard, LA
Belknap Co, NH
Columbiana Co, OH
Franklin Co, ME
Lafourche Parish, LA
Numerous Counties, MI
Numerous Counties, PA
Oxford Co, ME
St. James Parish, LA
St. Mary Parish, LA
Salem, OR
Somerset Co, ME
Victoria Co, TX

■ TABLE L-2 Carbon Monoxide Nonattainment Areas by Severity Classification (40 CFR 81.300 *et seq.*, as of July 9, 1994)

Serious

Los Angeles South Coast Air Basin, CA

Moderate $>$ 12.7 ppm

Anchorage, AK
Denver–Boulder, CO
Fresno, CA
Las Vegas, NV
New York–N. New Jersey–Long Island, NY–NJ–CT
Provo, UT
Seattle–Tacoma, WA
Spokane, WA

Moderate $\leq$ 12.7 ppm

Albuquerque, NM
Baltimore, MD
Boston, MA
Chico, CA
Cleveland, OH
Colorado Springs, CO
Duluth, —MN
El Paso, TX
Fairbanks, AK
Fort Collins, CO
Grants Pass, OR
Hartford–New Britain—Middletown, CT
Klamath Falls, OR
Lake Tahoe South Shore, CA
Longmont, CO
Medford, OR
Memphis, TN
Minneapolis–St. Paul, MN
Missoula, MT
Modesto, CA
Ogden, UT
Philadelphia–Camden Co, PA–NJ
Phoenix, AZ
Portland–Vancouver, OR–WA
Raleigh–Durham, NC
Reno, NV
Sacramento, CA
San Francisco–Oakland–San Jose, CA
San Diego, CA
Stockton, CA
Syracuse, NY
Washington, DC–MD–VA
Winston–Salem, NC

Unclassified

Atlantic City Area, NJ
Billings, MT
Boise, ID
Charlotte, NC
Detroit, MI
Eugene–Springfield, OR
Great Falls, MT
Manchester, NH
Middlesex Co, NJ
Nashua, NH
Pittsburgh, PA
St. Louis, MO
Salem, OR
Salt Lake City, UT

TABLE L-2 *(Continued)*

Greeley, CO	Trenton Area, NJ
Indianapolis, IN	Tucson, AZ
Lake Tahoe, NV	Yakima Co, WA

TABLE L-3 PM_{10} **Nonattainment Areas by State and Severity Classification (40 CFR 81.300 *et seq.*, as of July 9, 1994)**

Alabama
None

Alaska
Anchorage (moderate)
Juneau (moderate)

Arizona
Ajo (moderate)
Bullhead City (moderate)
Gilla County (part moderate)
Maricopa County (part moderate)
Nogales (moderate)
Paul Spur/Douglas (moderate)
Pima County (part moderate)
Pinal County (part moderate)
Yuma County (part moderate)

Arkansas
None

California
Coachella Valley (serious)
Imperial Valley (moderate)
Mono County (part moderate)
Owens Valley (serious)
San Joaquin Valley (serious)
Sacramento County (moderate)
Searles Valley (moderate)
South Coast Air Basin (serious)

Colorado
Aspen (moderate)
Canon City (moderate)
Denver–Boulder (moderate)
Lamar (moderate)
Pagosa Springs (moderate)
Routt County (part moderate)
Telluride (moderate)

Connecticut
City of New Haven (nonattainment)

Delaware
None

District of Columbia
None

Florida
None

Georgia
None

Hawaii
None

Idaho
Boise (moderate)
Pinehurst (moderate)
Pocatello (moderate)
Sandpoint (moderate)
Shoshone (part moderate)

■ **TABLE L-3** PM_{10} **Nonattainment Areas by State and Severity Classification (40 CFR 81.300 *et seq.*, as of July 9, 1994) *(Continued)***

Illinois
Cook County (part moderate)
La Salle County (part moderate)
Madison County (part moderate)

Indiana
Lake County (part moderate)
Vermillion County (part moderate)

Iowa
None

Kansas
None

Kentucky
None

Louisiana
None

Maine
Presque Isle (moderate)

Maryland
None

Massachusetts
None

Michigan
Wayne County (part moderate)

Minnesota
Olmstead County (part moderate)
Ramsey County (part moderate)

Mississippi
None

Missouri
None

Montana
Columbia Falls (moderate)
Flathead County (part moderate)
Lake County (part moderate)
Libby (moderate)
Missoula County (part moderate)
Rosebud County (part moderate)
Silver Bow County (part moderate)

Nebraska
None

Nevada
Las Vegas (serious)
Washoe County (part moderate)

New Hampshire
None

New Jersey
None

New Mexico
Dona Ana County (moderate)

New York
New York County (moderate)

North Carolina
None

North Dakota
None

Ohio
Cayahoga County (moderate)
Jefferson County (part moderate)

Oklahoma
None

■ **TABLE L-3** *(Continued)*

Oregon
Eugene–Springfield (moderate)
Grants Pass (moderate)
Jackson County (part moderate)
Klamath Falls (moderate)
LaGrande (moderate)

Pennsylvania
Allegheny County (part moderate)

Rhode Island
None

South Carolina
None

South Dakota
None

Tennessee
None

Texas
El Paso (moderate)

Utah
Salt Lake County (moderate)
Utah County (moderate)

Vermont
None

Virginia
None

Washington
King County (part moderate)
Pierce County (part moderate)
Spokane County (part moderate)
Thurston County (part moderate)
Walla Walla (part moderate)
Yakima County (part moderate)

West Virginia
Brooke (moderate)

Wisconsin
None

Wyoming
Sheridan (moderate)

■ **TABLE L-4** **Sulfur Dioxide Nonattainment Areas by State (40 CFR 81.300 *et seq.*, as of July 9, 1994)**

Alabama
None

Alaska
None

Arizona
Ajo (part)
Douglas (part)
Hayden (part)
Miami (part)
Morenci (part)
San Manuel (part)

Arkansas
None

California
None

Colorado
None

Connecticut
None

Delaware
None

■ **TABLE L-4 Sulfur Dioxide Nonattainment Areas by State (40 CFR 81.300 *et seq.*, as of July 9, 1994) *(Continued)***

District of Columbia
None

Florida
None

Georgia
None

Hawaii
None

Idaho
None

Illinois
Peoria County (part)
Tazewell County (part)

Indiana
Lake County (part)
LaPorte County (part)
Marion County (part)
Porter County (part)
Warrick County (part)
Wayne County (part)

Iowa
None

Kansas
None

Kentucky
Boyd County (part)
Muhlenberg County

Louisiana
None

Maine
Millinocket (part)

Maryland
Cumberland–Keyser AQCR

Massachusetts
None

Michigan
None

Minnesota
Rochester
AQCR 131

Mississippi
None

Missouri
None

Montana
Laurel
East Helena

Nebraska
None

Nevada
Steptoe Valley (part)

New Hampshire
None

New Jersey
Delaware Valley (part)

New Mexico
Grant County (part)

New York
None

North Carolina
None

North Dakota
None

■ **TABLE L-4** *(Continued)*

Ohio
Coshocton County (part)
Cuyahoga County (part)
Jefferson County (part)
Lake County (part)
Lorain County (part)
Lucas County (part)
Summit County (part)

Oklahoma
None

Oregon
None

Pennsylvania
Allegheny County (part)
Armstrong County (part)
Warren County (part)

Rhode Island
None

South Carolina
None

South Dakota
None

Tennessee
Benton County (part)
Humphreys County (part)
Polk County (part)

Texas
None

Utah
Salt Lake County (part)
Tooele County (part)

Vermont
None

Virginia
None

Washington
None

West Virginia
Steubenville–Weirton–Wheeling AQCR
New Monchester–Grant District

Wisconsin
Marathon County (part)
Oneida County (part)

Wyoming
None

■ **TABLE L-5** **Nitrogen Dioxide Nonattainment Areas by State (40 CFR 81.300 *et seq.*, as of July 9, 1994)**

Alabama
None

Alaska
None

Arizona
None

Arkansas
None

California
Los Angeles South Coast Air Basin

Colorado
None

■ TABLE L-5 Nitrogen Dioxide Nonattainment Areas by State (40 CFR 81.300 *et seq.*, as of July 9, 1994) *(Continued)*

Connecticut
None

Delaware
None

District of Columbia
None

Florida
None

Georgia
None

Hawaii
None

Idaho
None

Illinois
None

Indiana
None

Iowa
None

Kansas
None

Kentucky
None

Louisiana
None

Maine
None

Maryland
None

Massachusetts
None

Michigan
None

Minnesota
None

Mississippi
None

Missouri
None

Montana
None

Nebraska
None

Nevada
None

New Hampshire
None

New Jersey
None

New Mexico
None

New York
None

North Carolina
None

North Dakota
None

Ohio
None

Oklahoma
None

Oregon
None

Pennsylvania
None

Rhode Island
None

■ **TABLE L-5** ***(Continued)***

South Carolina	**Virginia**
None	None
South Dakota	**Washington**
None	None
Tennessee	**West Virginia**
Entire state	None
Texas	**Wisconsin**
None	None
Utah	**Wyoming**
None	None
Vermont	
None	

■ **TABLE L-6** **Lead Nonattainment Areas by State (40 CFR 81.300** ***et seq.*****, as of July 9, 1994)**

Alabama	**Georgia**
Jefferson County (part)	Muscogee County (part)
Alaska	**Hawaii**
None	None
Arizona	**Idaho**
None	None
Arkansas	**Illinois**
None	None
California	**Indiana**
None	Marion County (part)
Colorado	**Iowa**
None	None
Connecticut	**Kansas**
None	None
Delaware	**Kentucky**
None	None
District of Columbia	**Louisiana**
None	None
Florida	**Maine**
None	None

■ **TABLE L-6** **Lead Nonattainment Areas by State (40 CFR 81.300 *et seq.*, as of July 9, 1994) *(Continued)***

Maryland
None

Massachusetts
None

Michigan
None

Minnesota
Dakota County (part)

Mississippi
None

Missouri
Iron County (part)
Jefferson County (part)

Montana
East Helena

Nebraska
Omaha

Nevada
None

New Hampshire
None

New Jersey
None

New Mexico
None

New York
None

North Carolina
None

North Dakota
None

Ohio
None

Oklahoma
None

Oregon
None

Pennsylvania
None

Rhode Island
None

South Carolina
None

South Dakota
None

Tennessee
Fayette County (part)
Shelby County (part)
Williamson County (part)

Texas
Collin County (part)

Utah
None

Vermont
None

Virginia
None

Washington
None

West Virginia
None

Wisconsin
None

Wyoming
None

APPENDIX

M

EPA MEMORANDUM ON NONATTAINMENT REDESIGNATION REQUESTS

UNITED STATES ENVIRONMENTAL PROTECTION AGENCY
Office of Air Quality Planning and Standards
Research Triangle Park, North Carolina 27711

4 SEP 1992

MEMORANDUM

SUBJECT: Procedures for Processing Requests to Redesignate Areas to Attainment

FROM: John Calcagni, Director
Air Quality Management Division (MD-15)

TO: Director, Air, Pesticides and Toxics Management Division, Regions I and IV
Director, Air and Waste Management Division, Region II
Director, Air, Radiation and Toxics Division, Region III
Director, Air and Radiation Division, Region V
Director, Air, Pesticides and Toxics Division, Region VI
Director, Air and Toxics Division, Regions VII, VIII, IX, and X

Purpose

The Office of Air Quality Planning and Standards (OAQPS) expects that a number of redesignation requests will be submitted in the near future. Thus, Regions will need to have guidance on the applicable procedures for handling these requests, including maintenance plan provisions. This memorandum, therefore, consolidates the Environmental Protection Agency's (EPA's) guidance regarding the processing of requests for redesignation of nonattainment areas to attainment for ozone (O_3), carbon monoxide (CO), particulate matter (PM-10), sulfur dioxide (SO_2), nitrogen dioxide (NO_2), and lead (Pb). Regions should use this guidance as a general framework for drafting Federal Register notices pertaining to redesignation requests. Special concerns for areas seeking redesignation from unclassifiable to attainment will be addressed on a case-by-case basis.

Background

Section 107(d)(3)(E) of the Clean Air Act, as amended, states that an area can be redesignated to attainment if the following conditions are met:

1. The EPA has determined that the national ambient air quality standards (NAAQS) have been attained.

2. The applicable implementation plan has been fully approved by EPA under section 110(k).

3. The EPA has determined that the improvement in air quality is due to permanent and enforceable reductions in emissions.

4. The State has met all applicable requirements for the area under section 110 and Part D.

5. The EPA has fully approved a maintenance plan, including a contingency plan, for the area under section 175A.

Each of these criteria is discussed in more detail in the following paragraphs. Particular attention is given to maintenance plan provisions at the end of this document since maintenance plans constitute a new requirement under the amended Clean Air Act. Exceptions to the guidance will be considered on a case-by-case basis.

1. Attainment of the Standard

The State must show that the area is attaining the applicable NAAQS. There are two components involved in making this demonstration which should be considered interdependently. The first component relies upon ambient air quality data. The data that are used to demonstrate attainment should be the product of ambient monitoring that is representative of the area of highest concentration. These monitors should remain at the same location for the duration of the monitoring period required for demonstrating attainment. The data should be collected and quality-assured in accordance with 40 CFR 58 and recorded in the Aerometric Information Retrieval System (AIRS) in order for it to be available to the public for review. For purposes of redesignation, the Regional Office should verify that the integrity of the air quality monitoring network has been preserved.

For PM-10, an area may be considered attaining the NAAQS if the number of expected exceedances per year, according to 40 CFR 50.6, is less than or equal to 1.0. For O_3, the area must show that the average annual number of expected exceedances, according to 40 CFR 50.9, is less than or equal to 1.0 based on data from all monitoring sites in the area or its affected downwind environs. In making this showing, both PM-10 and O_3 must rely on 3 complete, consecutive calendar years of quality-assured air quality monitoring data, collected in accordance with 40 CFR 50, Appendices H and K. For CO, an area may be considered attaining the NAAQS if there are no violations, as determined in accordance with 40 CFR 50.8, based on 2 complete, consecutive calendar years of quality-assured monitoring data. For SO_2, according to 40 CFR 50.4, an area must show no more than one exceedance annually and for Pb, according to section 50.12, an area may show no exceedances on a quarterly basis.

The second component relies upon supplemental EPA-approved air quality modeling. No such supplemental modeling is required for O_3 nonattainment areas seeking redesignation. Modeling may be necessary to determine the representativeness of the monitored data. For pollutants such as SO_2 and CO, a small number of monitors typically is not representative of areawide air quality or areas of highest concentration. When dealing with SO_2, Pb, PM-10 (except for a limited number of initial moderate nonattainment areas), and CO (except moderate areas with design values of 12.7 parts per million or lower at the time of passage of the Clean Air Act Amendments of 1990), dispersion modeling will generally be necessary to evaluate comprehensively sources' impacts and to determine the areas of expected high concentrations based upon current conditions. Areas which were designated nonattainment based on modeling will generally not be redesignated to attainment unless an acceptable modeling analysis indicates attainment. Regions should consult with OAQPS for further guidance addressing the need for modeling in specific circumstances.

2. <u>State Implementation Plan (SIP) Approval</u>

The SIP for the area must be fully approved under section 110(k),[1] and must satisfy all requirements that apply to the area. It should be noted that approval action on SIP elements and the redesignation request may occur simultaneously. An area cannot be redesignated if a required element of its plan is the subject of a disapproval; a finding of failure to submit or to implement the SIP; or partial, conditional, or limited approval. However, this does not mean that earlier issues with regard to the SIP will be reopened. Regions should not reconsider those things that have already been approved and for which the Clean Air Act Amendments did not alter what is required. In contrast, to the extent the Amendments add a requirement or alter an existing requirement so that it adds something more, Regions should consider those issues. In addition, requests from areas known to be affected by dispersion techniques which are inconsistent with EPA guidance will continue to be considered unapprovable under section 110 and will not qualify for redesignation.

---

[1]Section 110(k) contains the requirements for EPA action on plan submissions. It addresses completeness, deadlines, full and partial approval, conditional approval, and disapproval.

3. Permanent and Enforceable Improvement in Air Quality

The State must be able to reasonably attribute the improvement in air quality to emission reductions which are permanent and enforceable.[2] Attainment resulting from temporary reductions in emission rates (e.g., reduced production or shutdown due to temporary adverse economic conditions) or unusually favorable meteorology would not qualify as an air quality improvement due to permanent and enforceable emission reductions.

In making this showing, the State should estimate the percent reduction (from the year that was used to determine the design value for designation and classification) achieved from Federal measures such as the Federal Motor Vehicle Control Program and fuel volatility rules as well as control measures that have been adopted and implemented by the State. This estimate should consider emission rates, production capacities, and other related information to clearly show that the air quality improvements are the result of implemented controls. The analysis should assume that sources are operating at permitted levels (or historic peak levels) unless evidence is presented that such an assumption is unrealistic.

4. Section 110 and Part D Requirements

For the purposes of redesignation, a State must meet all requirements of section 110 and Part D that were applicable prior to submittal of the complete redesignation request. When evaluating a redesignation request, Regions should not consider whether the State has met requirements that come due under the Act after submittal of a complete redesignation request.[3]

---

[2]This is consistent with EPA's existing policy on redesignations as stated in an April 21, 1983 memorandum titled "Section 107 Designation Policy Summary." This memorandum states that in order for an area to be redesignated to attainment, the State must show that "actual enforceable emission reductions are responsible for the recent air quality improvement." This element of the policy retains its validity under the amended Act pursuant to section 193. [Note: other aspects of the April 21, 1983 memorandum have since been superseded by subsequent memorandums; interested parties should consult with OAQPS before relying on these aspects, e.g. those relating to required years of air quality data.]

[3]Under section 175A(c), however, the requirements of Part D remain in force and effect for the area until such time as it is redesignated. Upon redesignation to attainment, the requirements that became due under section 175A(c) after submittal of the complete redesignation request would no longer be applicable.

However, any requirements that came due prior to submittal of the redesignation request must be fully approved into the plan at or before the time EPA redesignates the area.

To avoid confusion concerning what requirements will be applicable for purposes of redesignation, Regions should encourage States to work closely with the appropriate Regional Office early in the process. This will help to ensure that a redesignation request submitted by the State has a high likelihood of being approved by EPA. Regions should advise States of the practical planning consequences if EPA disapproves the redesignation request or if the request is invalidated because of violations recorded during EPA's review. Under such circumstances, EPA does not have the discretion to adjust schedules for implementing SIP requirements. As a result, an area may risk sanctions and/or Federal implementation plan implementation that could result from failure to meet SIP submittal or implementation requirements.

a. Section 110 Requirements

Section 110(a)(2) contains general requirements for nonattainment plans. Most of the provisions of this section are the same as those contained in the pre-amended Act. We will provide guidance on these requirements as needed.[4]

b. Part D Requirements

Part D consists of general requirements applicable to all areas which are designated nonattainment based on a violation of the NAAQS. The general requirements are followed by a series of subparts specific to each pollutant. The general requirements appear in subpart 1. The requirements relating to O_3, CO, PM-10, SO_2, NO_2, and Pb appear in subparts 2 through 5. In those instances where an area is subject to both the general nonattainment provisions in subpart 1 as well as one of the pollutant-specific subparts, the general provisions may be subsumed within, or superseded by, the more specific requirements of subparts 2 through 5.

If an area was not classified under section 181 for O_3, or section 186 for CO, then that area is only subject to the provisions of subpart 1, "Nonattainment Areas in General." In addition to relevant provisions in subpart 1, an O_3 and CO area, which is classified, must meet all applicable requirements in subpart 2, "Additional Provisions for Ozone Nonattainment Areas," and subpart 3, "Additional Provisions for Carbon Monoxide

---

[4]General guidance regarding the requirements for SIP's may be found in the "General Preamble to Title I of the 1990 Clean Air Act Amendments," 57 FR 13498 (April 16, 1992).

Nonattainment Areas," respectively, before the area may be redesignated to attainment. All PM-10 nonattainment areas (whether classified as moderate or serious) must similarly meet the applicable general provisions of subpart 1 and the specific PM-10 provisions in subpart 4, "Additional Provisions for Particulate Matter Nonattainment Areas." Likewise, SO_2, NO_2, and Pb nonattainment areas are subject to the applicable general nonattainment provisions in subpart 1 as well as the more specific requirements in subpart 5, "Additional Provisions for Areas Designated Nonattainment for Sulfur Oxides, Nitrogen Dioxide, and Lead."

i. <u>Section 172(c) Requirements</u>

This section contains general requirements for nonattainment plans. A thorough discussion of these requirements may be found in the <u>General Preamble to Title I</u> [57 FR 13498 (April 16, 1992)]. The EPA anticipates that areas will already have met most or all of these requirements to the extent that they are not superseded by more specific Part D requirements. The requirements for reasonable further progress, identification of certain emissions increases, and other measures needed for attainment will not apply for redesignations because they only have meaning for areas not attaining the standard. The requirements for an emission inventory will be satisfied by the inventory requirements of the maintenance plan. The requirements of the Part D new source review program will be replaced by the prevention of significant deterioration (PSD) program once the area has been redesignated. However, in order to ensure that the PSD program will become fully effective immediately upon redesignation, either the State must be delegated the Federal PSD program or the State must make any needed modifications to its rules to have the approved PSD program apply to the affected area upon redesignation.

ii. <u>Conformity</u>

The State must work with EPA to show that its SIP provisions are consistent with section 176(c)(4) conformity requirements. The redesignation request should include conformity procedures, if the State already has these procedures in place. Additionally, we currently interpret the conformity requirement to apply to attainment areas. However, EPA has not yet issued its conformity regulations specifying what areas are subject to the conformity requirement. Therefore, if a State does not have conformity procedures in place at the time that it submits a redesignation request, the State must commit to follow EPA's conformity regulation upon issuance, as applicable. If the State submits the redesignation request subsequent to EPA's issuance of the conformity regulations, and the conformity requirement became applicable to the area prior to submission,

7

the State must adopt the applicable conformity requirements before EPA can redesignate the area.

5. Maintenance Plans

Section 107(d)(3)(E) of the amended Act stipulates that for an area to be redesignated, EPA must fully approve a maintenance plan which meets the requirements of section 175A. A State may submit both the redesignation request and the maintenance plan at the same time and rulemaking on both may proceed on a parallel track. Maintenance plans may, of course, be submitted and approved by EPA before a redesignation is requested. However, according to section 175A(c), pending approval of the maintenance plan and redesignation request, all applicable nonattainment area requirements shall remain in place.

Section 175A defines the general framework of a maintenance plan. The maintenance plan will constitute a SIP revision and must provide for maintenance of the relevant NAAQS in the area for at least 10 years after redesignation. Section 175A further states that the plan shall contain such additional measures, if any, as may be necessary to ensure such maintenance. Because the Act requires a demonstration of maintenance for 10 years after an area is redesignated (not 10 years after submittal of a redesignation request), the State should plan for some lead time for EPA action on the request. In other words, the maintenance demonstration should project maintenance for 10 years, beginning from a date which factors in the time necessary for EPA review and approval action on the redesignation request. In determining the amount of lead time to allow, States should consider that section 107(d)(3)(D) grants the Administrator up to 18 months from receipt of a complete submittal to process a redesignation request. The statute also requires the State to submit a revision of the SIP 8 years after the original redesignation request is approved to provide for maintenance of the NAAQS for an additional 10 years following the first 10-year period [see section 175A(b)].

In addition, the maintenance plan shall contain such contingency measures as the Administrator deems necessary to ensure prompt correction of any violation of the NAAQS [see section 175A(d)]. The Act provides that, at a minimum, the contingency measures must include a requirement that the State will implement all measures contained in the nonattainment SIP prior to redesignation. Failure to maintain the NAAQS and triggering of the contingency plan will not necessitate a revision of the SIP unless required by the Administrator, as stated in section 175A(d).

The following is a list of core provisions that we anticipate will be necessary to ensure maintenance of the relevant NAAQS in an area seeking redesignation from nonattainment to attainment. We therefore recommend that States seeking redesignation of a nonattainment area consider these provisions. However, any final EPA determination regarding the adequacy of a maintenance plan will be made following review of the plan submittal in light of the particular circumstances facing the area proposed for redesignation and based on all relevant information available at the time.

a. Attainment Inventory

The State should develop an attainment emissions inventory to identify the level of emissions in the area which is sufficient to attain the NAAQS.[5] This inventory should be consistent with EPA's most recent guidance on emission inventories for nonattainment areas available at the time and should include the emissions during the time period associated with the monitoring data showing attainment.[6]

Source size thresholds are 100 tons/year for SO_2, NO_2, and PM-10 areas, and 5 tons/year for Pb based upon 40 CFR 51.100(k) and 51.322, as well as established practice for AIRS data. The source size threshold for serious PM-10 areas is 70 tons/year

---

[5]Where the State has made an adequate demonstration that air quality has improved as a result of the SIP (as discussed previously), the attainment inventory will generally be the actual inventory at the time the area attained the standard.

[6]The EPA's current guidance on the preparation of emission inventories for O_3 and CO nonattainment areas is contained in the following documents: "Procedures for the Preparation of Emission Inventories for Carbon Monoxide and Precursors of Ozone: Volume I" (EPA-450/4-91-016), "Procedures for the Preparation of Emission Inventories for Carbon Monoxide and Precursors of Ozone: Volume II" (EPA-450/4-91-014), "Emission Inventory Requirements for Ozone State Implementation Plans" (EPA-450/4-91-010), "Emission Inventory Requirements for Carbon Monoxide Implementation Plans" (EPA-450/4-91-011), "Guideline for Regulatory Application of the Urban Airshed Model" (EPA-450/4-91-013), "Procedures for Emission Inventory Preparation: Volume IV, Mobile Sources" (EPA-450/4-81-026d), and "Procedures for Preparing Emission Inventory Projections" (EPA-450/4-91-019). The EPA does not currently have specific guidance on attainment emissions inventories for SO_2. In lieu thereof, States are referred to the guidance on emissions data to be used as input to modeling demonstrations, contained in Table 9.1 of EPA's "Guideline on Air Quality Models (Revised)" (EPA-450/2-78-027R), July 1987, which is generally applicable to all criteria pollutants. Emission inventory procedures and requirements documents are currently being prepared by OAQPS for PM-10 and Pb; these documents are due for release by summer 1992.

according to Clean Air Act section 189(b)(3). However, the inventory should include sources below these size thresholds if these smaller sources were included in the SIP attainment demonstration. Where sources below the 100, 70, and 5 tons/year-size thresholds (e.g., areas with smaller source size definitions) are subject to a State's minor source permit program, these sources need only be addressed in the aggregate to the extent that they result in areawide growth.

For O_3 nonattainment areas, the inventory should be based on actual "typical summer day" emissions of O_3 precursors (volatile organic compounds and nitrogen oxides) during the attainment year. This will generally correspond to one of the periodic inventories required for nonattainment areas to reconcile milestones. For CO nonattainment areas, the inventory should be based on actual "typical CO season day" emissions for the attainment year. This will generally correspond to one of the periodic inventories required for nonattainment areas.

b. <u>Maintenance Demonstration</u>

A State may generally demonstrate maintenance of the NAAQS by either showing that future emissions of a pollutant or its precursors will not exceed the level of the attainment inventory, or by modeling to show that the future mix of sources and emission rates will not cause a violation of the NAAQS. Under the Clean Air Act, many areas are required to submit modeled attainment demonstrations to show that proposed reductions in emissions will be sufficient to attain the applicable NAAQS. For these areas, the maintenance demonstration should be based upon the same level of modeling. In areas where no such modeling was required, the State should be able to rely on the attainment inventory approach. In both instances, the demonstration should be for a period of 10 years following the redesignation.

Where modeling is relied upon to demonstrate maintenance, each plan should contain a summary of the air quality concentrations expected to result from application of the control strategy. In the process, the plan should identify and describe the dispersion model or other air quality model used to project ambient concentrations (see 40 CFR 51.46).

In either case, to satisfy the demonstration requirement the State should project emissions for the 10-year period following redesignation, either for the purpose of showing that emissions will not increase over the attainment inventory or for conducting modeling. The projected inventory should consider future growth, including population and industry, should be consistent

---

[7]Guidance for projecting emissions may be found in the emissions inventory guidance cited in footnote 6.

with the attainment inventory, and should document data inputs and assumptions. All elements of the demonstration (e.g., emission projections, new source growth, and modeling) should be consistent with current EPA modeling guidance.[8] For O_3 and CO, the projected emissions should reflect the expected actual emissions based on enforceable emission rates and typical production rates.

For CO, a State should address the areawide component of the maintenance demonstration either by showing that future CO emissions will not increase or by conducting areawide modeling. Preferably, the State should carry out hot-spot modeling that is consistent with the Guideline on Air Quality Models (Revised), in order to demonstrate maintenance of the NAAQS. In particular, if the nonattainment problem is related to a pattern of hot-spots then hot-spot modeling should generally be conducted. However, hot-spot modeling is not automatically required. For example, if the nonattainment problem was related solely to stationary point sources, or if highway improvements have been implemented and the associated emission reductions and travel characteristics can be qualitatively documented, then hot-spot modeling is not required. In such cases, adequate documentation as well as the concurrence of Headquarters is needed.

Any assumptions concerning emission rates must reflect permanent, enforceable measures. In other words, a State generally cannot take credit in the maintenance demonstration for reductions unless there are regulations in place requiring those reductions or the reductions are otherwise shown to be permanent. Therefore, the State will be expected to maintain its implemented control strategy despite redesignation to attainment, unless such measures are shown to be unnecessary for maintenance or are replaced with measures that achieve equivalent reductions (see additional discussion under "Contingency Plan"). Emission reductions from source shutdowns can be considered permanent and enforceable to the extent that those shutdowns have been reflected in the SIP and all applicable permits have been modified accordingly.

Modeling used to demonstrate attainment may be relied upon in the maintenance demonstration where the modeling conforms to current EPA guidance and where the State has projected no significant changes in the modeling inputs during the intervening time. Where the original attainment demonstration may no longer be relied upon, States will be expected to remodel using current

---

[8]The EPA-approved modeling guidance may be found in the following documents: "Guideline on Air Quality Models (Revised)," OAQPS, RTP, NC (EPA-450/2-78-027R), July 1986; and "PM-10 SIP Development Guideline," OAQPS, RTP, NC (EPA-450/2-86-001), June 1987.

EPA referenced techniques.[9] This may be necessary where, for example, there has been a change in emissions or a change in the siting of new sources or modifications such that air quality may no longer be accurately represented by the existing modeling.

c. Monitoring Network

Once an area has been redesignated, the State should continue to operate an appropriate air quality monitoring network, in accordance with 40 CFR Part 58, to verify the attainment status of the area. The maintenance plan should contain provisions for continued operation of air quality monitors that will provide such verification. In cases where measured mobile source parameters (e.g., vehicle miles traveled congestion) have changed over time, the State may also need to perform a saturation monitoring study to determine the need for, and location of, additional permanent monitors.

d. Verification of Continued Attainment

Each State should ensure that it has the legal authority to implement and enforce all measures necessary to attain and to maintain the NAAQS. Sections 110(a)(2)(B) and (F) of the Clean Air Act, as amended, and regulations promulgated at 40 CFR 51.110(k), suggest that one such measure is the acquisition of ambient and source emission data to demonstrate attainment and maintenance.

Regardless of whether the maintenance demonstration is based on a showing that future emission inventories will not exceed the attainment inventory or on modeling, the State submittal should indicate how the State will track the progress of the maintenance plan. This is necessary due to the fact that the emission projections made for the maintenance demonstration depend on assumptions of point and area source growth.

One option for tracking the progress of the maintenance demonstration, provided here as an example, would be for the State to periodically update the emissions inventory. In this case, the maintenance plan should specify the frequency of any planned inventory updates. Such an update could be based, in part, on the annual AIRS update and could indicate new source growth and other changes from the attainment inventory (e.g., changes in vehicle miles travelled or in traffic patterns). As an alternative to a complete update of the inventory, the State may choose to do a comprehensive review of the factors that were used in developing the attainment inventory to show no significant change. If this review does show a significant change, the State should then perform an update of the inventory.

---

[9]See references for modeling guidance cited in footnote 8.

Where the demonstration is based on modeling, an option for tracking progress would be for the State to periodically (typically every 3 years) reevaluate the modeling assumptions and input data. In any event, the State should monitor the indicators for triggering contingency measures (as discussed below).

e. Contingency Plan

Section 175A of the Act also requires that a maintenance plan include contingency provisions, as necessary, to promptly correct any violation of the NAAQS that occurs after redesignation of the area. These contingency measures are distinguished from those generally required for nonattainment areas under section 172(c)(9) and those specifically required for O_3 and CO nonattainment areas under sections 182(c)(9) and 187(a)(3), respectively. For the purposes of section 175A, a State is not required to have fully adopted contingency measures that will take effect without further action by the State in order for the maintenance plan to be approved. However, the contingency plan is considered to be an enforceable part of the SIP and should ensure that the contingency measures are adopted expediently once they are triggered. The plan should clearly identify the measures to be adopted, a schedule and procedure for adoption and implementation, and a specific time limit for action by the State. As a necessary part of the plan, the State should also identify specific indicators, or triggers, which will be used to determine when the contingency measures need to be implemented.

Where the maintenance demonstration is based on the inventory, the State may, for example, identify an "action level" of emissions as the indicator. If later inventory updates show that the inventory has exceeded the action level, the State would take the necessary steps to implement the contingency measures. The indicators would allow a State to take early action to address potential violations of the NAAQS before they occur. By taking early action, States may be able to prevent any actual violations of the NAAQS and, therefore, eliminate the need on the part of EPA to redesignate an area to nonattainment.

Other indicators to consider include monitored or modeled violations of the NAAQS (due to the inadequacy of monitoring data in some situations). It is important to note that air quality data in excess of the NAAQS will not automatically necessitate a revision of the SIP where implementation of contingency measures is adequate to address the cause of the violation. The need for a SIP revision is subject to the Administrator's discretion.

The EPA will review what constitutes a contingency plan on a case-by-case basis. At a minimum, it must require that the State will implement all measures contained in the Part D nonattainment plan for the area prior to redesignation [see section 175A(d)]. This language suggests that a State may submit a SIP revision at the time of its redesignation request to remove or reduce the stringency of control measures. Such a revision can be approved by EPA if it provides for compensating equivalent reductions. A demonstration that measures are equivalent would have to include appropriate modeling or an adequate justification. Alternatively, a State might be able to demonstrate (through EPA-approved modeling) that the measures are not necessary for maintenance of the standard. In either case, the contingency plan would have to provide for implementation of any measures that were reduced or removed after redesignation of the area.

Summary

As stated previously, this memorandum consolidates EPA's redesignation and maintenance plan guidance and Regions should rely upon it as a general framework in drafting Federal Register notices. It is strongly suggested that the Regional Offices share this document with the appropriate States. This should give the States a better understanding of what is expected from a redesignation request and maintenance plan under existing policy. Any necessary changes to existing Agency policy will be made through our action on specific redesignation requests and the review of section 175A maintenance plans for these particular areas, both of which are subject to notice and comment rulemaking procedures. Thus, in applying this memorandum to specific circumstances in a rulemaking, Regions should consider the applicability of the underlying policies to the particular facts and to comments submitted by any person. If your staff members have questions which require clarification, they may contact Sharon Reinders at (919) 541-5284 for O_3- and CO-related issues, and Eric Ginsburg at (919) 541-0877 for SO_2-, PM-10-, and Pb-related issues.

cc: Chief, Air Branch, Regions I-X
John Cabaniss, OMS
Denise Devoe, OAQPS
Bill Laxton, TSD
Rich Ossias, OGC
John Rasnic, SSCD
John Seitz, OAQPS
Mike Shapiro, OAR
Lydia Wegman, OAQPS

APPENDIX

N

DRAFT MODEL TITLE V PERMIT FOR GRAPHIC ARTS

Note: This document is a manual reproduction of material obtained from EPA.

MODEL PERMIT FOR GRAPHIC ARTS

SECTION E: SPECIFIC PERMIT CONDITIONS

[Applicability

(1) This permit applies to any packaging rotogravure, publication rotogravure, or flexographic printing press at any facility whose maximum theoretical emissions of volatile organic compounds (VOCs) (including solvents used to clean each of these printing presses) without control devices from all printing presses are greater than or equal to 90.7 megagrams (Mg) (100 tons) per year.

(2) An owner or operator whose emissions are below the applicability threshold in (1) above shall comply with any certification, recordkeeping, and reporting requirements for exempt sources.

(3) Any facility that becomes or is currently subject to the requirements in this permit by exceeding the applicability threshold in (1) above will remain subject to these requirements even if its emissions later fall below the applicability threshold.

(4) Any facility that is currently subject to a State or Federal rule promulgated pursuant to the Clean Air Act Amendments of 1977 by exceeding an applicability threshold is and will remain subject to these provisions, even if its throughput or emissions have fallen or later fall below the applicability threshold.]

1. Definitions [§§XX.3000, XX.3035(b)]

[Insert applicable definitions.]

2. Emission Limits, Standards, and Other Emissions-Related Requirements

[§ XX.3035(c)–(e)]

****

[Use of Complying Coatings or Inks]

a. Packaging rotogravure or flexographic printing presses. The permittee shall not apply any coating or ink unless the VOC content is equal to or less than:

i. 40 percent VOC by volume of the coating or ink, excluding water, as applied;

ii. 25 percent VOC by volume of the volatile content in the coating or ink, as applied; or

iii. 0.5 kilogram (kg) VOC per kg (0.5 pound [lb] VOC per lb) coating solids, as applied.

b. Publication rotogravure printing presses. The permittee shall not apply any coating or ink unless the VOC content is equal to or less than:

i. 40 percent VOC by volume of the coating or ink, excluding water, as applied; or

ii. 25 percent VOC by volume of the volatile content in the coating or ink, as applied.

****

[Use of Daily-Weighted Averaging]

a. The permittee shall not apply coatings or inks on the permitted printing press unless the daily-weighted average, by volume, VOC content of all coatings and inks, as applied, each day on the permitted printing press is equal to or less than:

i. 40 percent VOC by volume of the coating or ink, excluding water, as applied, as determined using the following equation:

$$VOC_{(A)} = \frac{\sum_{i=1}^{n} L_i V_{VOCi}}{\sum_{i=1}^{n} (V_{si} + V_{VOCi})} \times 100$$

where

$VOC_{(A)}$ = the weighted average VOC content in units of percent VOC by volume of all coatings and inks (excluding water and exempt compounds) used each day;

i = subscript denoting a specific coating or ink, as applied;

n = the number of different coatings and/or inks, as applied, each day on a printing press;

L_i = the liquid volume of each coating or ink, as applied, used that day in units of liters (L) (gallons [gal]);

V_{si} = the volume fraction of solids in each coating or ink, as applied; and

V_{VOCi} = the volume fraction of VOC in each coating or ink, as applied; or

ii. 25 percent VOC by volume of the volatile content in the coating or ink, as applied, as determined using the following equation:

$$VOC_{(B)} = \frac{\sum_{i=1}^{n} L_i V_{VOCi}}{\sum_{i=1}^{n} L_i V_{VCi}} \times 100$$

where

$VOC_{(B)}$ = the weighted average VOC content in units of percent VOC by volume of the volatile content of all coatings and inks used each day; and

V_{VCi} = the volume fraction of volatile matter in each coating or ink, as applied; or

iii. For packaging rotogravure or flexographic printing presses, 0.5 kilogram (kg) VOC per kg (0.5 pound [lb] VOC per lb) coating solids, as applied, as determined using the following equation:

$$VOC_{(C)} = \frac{\sum_{i=1}^{n} L_i D_i W_{VOCi}}{\sum_{i=1}^{n} L_i D_i W_{si}} \times 100$$

where

$VOC_{(C)}$ = the weighted average VOC content in units of mass of VOC per mass of coating solids;

D_i = the density of each coating or ink, as applied, in units of mass of coating or ink per unit volume of coating or ink;

W_{VOCi} = the weight fraction of VOC in each coating or ink, as applied; and

W_{si} = the weight fraction of solids in each coating or ink, as applied.

b. The permittee may comply with the daily-weighted average limitation by grouping coatings or inks used on a printing press into two categories that meet the following conditions:

i. The daily-weighted average VOC content for the first category must comply with (a) (i) of this condition, as determined by applying the equation in (a) (i) of this condition to the coatings or inks in this first category.

ii. The daily-weighted average VOC content for the second category must comply with (a) (ii) of this condition, as determined by applying the equation in (a) (ii) of this condition to the coatings or inks in this second category.

c. For packaging rotogravure or flexographic printing presses, any use of averaging between the two categories of coatings or inks requires compliance with the emission standard in (a) (iii) of this condition, as determined by applying the equation in (a) (iii) of this condition.

****

[Use of Control Devices]

a. The permittee shall:

i. Use a carbon adsorption control device that reduces VOC emissions delivered from the capture system to the control device by at least 90 percent, by weight;

ii. Use an incineration control device to reduce VOC emissions delivered from the capture system to the control device by at least 90 percent, by weight; or

iii. Use any other VOC emission control device to reduce VOC emissions delivered from the capture system to the control device by at least 90 percent, by weight.

b. The permittee shall equip the printing press with a capture system and control device that provides an overall emission reduction efficiency of at least [**insert 75 percent for a publication rotogravure printing press, 65 percent for a packaging rotogravure printing press, and/or 60 percent for a flexographic printing press**].

c. The permittee shall ensure that:

i. A capture system and control device are operated at all times that the printing press is in operation, and the permittee demonstrates compliance with this permit through the applicable coating analysis and capture system and control device efficiency test methods specified in Attachments 2 and 4 through 7 to this permit and in accordance with the capture efficiency test methods in 40 CFR Part 60, Appendix A [to be published]; and

ii. The control device is equipped with the applicable monitoring equipment specified in (b) of Attachment 4 to this permit, and the monitoring equipment is installed, calibrated, operated, and maintained according to the vendor's specifications at all times the control device is in use.

****

3. Handling, Storage, and Disposal of VOCs [§XX.3006]

The permittee shall not:

a. Cause, allow, or permit the disposal of more than 5 kg (11 lb) of any VOC, or of any materials containing more than 5 kg (11 lb) of any VOCs, at the facility in any 1 day in a manner that would permit the VOC to evaporate into the ambient air. This includes, but is not limited to, the disposal of VOC from VOC control devices;

b. Use open containers to store or dispose of cloth or paper impregnated with VOCs that are used for surface preparation, cleanup, or coating removal;

c. Store in open containers spent or fresh VOC to be used for surface preparation, cleanup, or coating removal; or

d. Use VOC to clean up spray equipment unless equipment is used to collect the cleaning compounds and to minimize their evaporation to the atmosphere.

4. Schedule of Compliance: [Memorandum: J. Calcagni, EPA: AQMD, and J. Seitz, EPA:SSCD, to D. Kee, EPA:Region V. March 16, 1989. Compliance Schedules for Volatile Organic Compounds.]

The permittee must be in compliance on or before [**insert the date of issuance of this permit; or insert a date up to 1 year from the effective date of the rule if either:**

a. The source was previously exempted and is newly regulated; or

b. **The source complied with the previously approved SIP but now requires major process changes and/or major capital expenditures to comply with the revised rule**].

5. Initial Compliance Certification [§XX.3035(g)(2)–(4)(i)]

****

[Use of Complying Coatings or Inks]

By [**insert date 1 year after date of promulgation of final rule**], the permittee shall certify to the permitting authority (PA) that the printing press will be in compliance with the emission limits in Condition 2 of this permit on and after the latter of [**insert date 1 year after date of promulgation of final rule**] or the initial startup date. Such certification shall include:

a. The name and location of the facility;
b. The address and telephone number of the person responsible for the facility;
c. Identification of the subject printing press(es);
d. The name and identification number of each coating and ink, as applied, each day; and
e. The VOC content of each coating and ink, as applied, each day.

****

[Use of Daily-Weighted Averaging]

By [**insert date 1 year after date of promulgation of final rule**], the permittee shall certify to the permitting authority (PA) that the printing press will comply with the emission limits in Condition 2 of this permit on and after the latter of [**insert date 1 year after date of promulgation of final rule**] or the initial startup date. Such certification shall include:

a. The name and location of the facility;
b. The address and telephone number of the person responsible for the facility;
c. The name and identification of each printing press that will comply by means of daily-weighted averaging;
d. The name and identification number of each coating and ink, as applied, each day available for use on each printing press;

e. The VOC content of each coating and ink, as applied, each day on each printing press expressed in units necessary to determine compliance;
f. The instrument or method by which the permittee will accurately measure or calculate the volume of each coating and ink, as applied, each day on each printing press;
g. The method by which the permittee will create and maintain records each day as required in Condition 8 of this permit;
h. An example of the format in which the records required in Condition 8 of this permit will be kept; and
i. The time at which the facility's "day" begins if a time other than midnight local time is used to define a "day."

****

[Use of Control Devices]

By [**insert date 1 year after date of promulgation of final rule**], the permittee shall perform all tests and submit to the permitting authority (PA) the results of all tests and calculations necessary to demonstrate that the permitted printing press will be in compliance with Condition 2 of this permit on and after the latter of [**insert date 1 year after date of promulgation of final rule**] or the initial startup date.

****

6. Inspection Requirements

[Insert inspection requirements that are consistent with existing State or Federal regulations.]

7. Monitoring Requirements [§XX.3083(b)(2)]

****

[Use of Complying Coatings or Inks or Daily-Weighted Averaging]

[Insert monitoring requirements that are consistent with existing State or Federal regulations.]

****

[Use of Control Devices]

See (b)(2) of Attachment 4 to this permit.

****

8. Recordkeeping Requirements [40 CFR 70.6(a)(3), §XX.3004, §XX.3035(g)(2)–(4)(ii)]

a. The permittee shall record:

- **i.** Changes at the permitted facility that cause a regulated pollutant not regulated by this permit to be emitted; and
- **ii.** The emissions resulting from the change.

b. The permittee shall maintain all records necessary for determining compliance with this permit [**specify location of these records, i.e. at the site or at an accessible location, and if the records are stored offsite, they should be made available upon request**] for a period of 5 years.

****

[Use of Complying Coatings or Inks]

c. Permittees complying with Condition 2(a)(i) or (b)(i) of this permit shall collect and record all of the following information each day for each printing press and maintain the information [**specify location of this information, i.e. at the site or at an accessible location, and if the information is stored offsite, it should be made available upon request**] for a period of 5 years:

- **i.** The name and identification number of each coating and ink, as applied, each day on each printing press; and
- **ii.** The VOC content of each coating and ink, as applied, each day on each printing press, expressed in units necessary to determine compliance.

****

[Use of Daily-Weighted Averaging]

c. The permittee shall collect and record all of the following information each day for each printing press and maintain the information [**specify location of the information, i.e. at the site or at an accessible loca-**

tion, and if the information is stored offsite, it should be made available upon request] for a period of 5 years:

i. The name and identification number of each coating and ink, as applied, each day on each printing press;

ii. The VOC content and the volume of each coating and ink, as applied, each day on each printing press, expressed in units necessary to determine compliance; and

iii. The daily-weighted average VOC content of all coatings and inks, as applied, each day on each printing press.

****

[Use of Control Devices]

c. The permittee shall collect and record all of the following information each day for each printing press and maintain the information [**specify location of the information, i.e. at the site or at an accessible location, and if the information is stored offsite, it should be made available upon request**] for a period of 5 years:

i. Control device monitoring data;

ii. A log of operating time for the capture system, control device, and monitoring equipment and the associated printing press;

iii. A maintenance log for the capture system, control device, and monitoring equipment detailing all routine and nonroutine maintenance performed including dates and duration of any outages.

****

9. Reporting Requirements [40 CFR 70.6(a)(3), §XX.3035 (g)(2)–(4)(iii)]

The permittee shall:

a. Report to the PA [**specify time period and whether by telephone or by mail**] deviations from the requirements of this permit, including those attributable to malfunctions as defined in 40 CFR 60.2. The report shall describe the cause of such deviations and any corrective actions or preventive measures taken;

b. Submit records of monitoring required in Condition 7 of this permit to the PA at least every 6 months and shall clearly identify all instances of deviations from the requirements of this permit;

c. Send [**insert desired number**] copies of all required reports to: [**insert the PA field office address**];

d. Have a responsible corporate official certify as being true, accurate, and complete any document, including reports, required to be submitted by this permit;

e. Report any record showing noncompliance with Condition 2 of this permit to the PA within [**the permitting agency should specify a time period here that is consistent with State regulations. It is suggested that the specified time period does not exceed 30 days**] following the occurrence; and

f. Notify the PA at least 30 calendar days before changing the method of compliance between daily-weighted averaging, use of complying coatings or inks, and control devices.

10. Test Methods and Compliance Procedures [§XX.3035(f)]

The permittee shall use the applicable test methods and compliance procedures specified in Attachments 1 through 7 to this permit to determine the VOC content of each coating and ink and the efficiency of any capture systems and control devices and to establish the records required under Condition 8 of this permit.

APPENDIX

O

DRAFT MODEL TITLE V PERMIT FOR PM_{10} EMISSIONS FROM LIME MANUFACTURING PLANTS

Note: This document is a manual reproduction of material obtained from EPA.

Model Operating Permit for PM_{10} Emissions from Lime Manufacturing Plants

Draft Final Report

March 1993

Prepared by

TRC Environmental Corporation
100 Europa Drive, Suite 150
Chapel Hill, NC 27514

EPA Contract No. 68-D0-0121
Work Assignment No. II-91

EPA Work Assignment Manager: W. Stephen Fruh
Emission Standards Division
Office of Air Quality Planning and Standards
U.S. Environmental Protection Agency
Research Triangle Park, NC 27711

This document was obtained from the EPA Office of Air Quality Planning and Standards Bulletin Board. It is publically available; however, it has been marked by EPA as follows:

This document has not been peer or administratively reviewed within EPA and is for agency use/distribution only. Do not quote, cite or distribute.

DISCLAIMER

This Draft Final Report was prepared for the Emission Standards Division, U.S. Environmental Protection Agency, Research Triangle Park, North Carolina by TRC Environmental Corporation, 100 Europa Drive, Chapel Hill, North Carolina 27514, in partial fulfillment of Contract No. 68-D0-0121, Work Assignment No. II-91. The opinions, findings, and conclusions expressed are those of the authors and not necessarily those of the Environmental Protection Agency.

TABLE OF CONTENTS

LIST OF FIGURES

LIST OF TABLES

1.0 INTRODUCTION

Title V of the 1990 Clean Air Act Amendments (CAAA) requires States to adopt operating permit programs to be administered by State or local air pollution control agencies. The programs are to issue operating permits to major stationary sources of criteria and hazardous air pollutants, sources covered by new source performance standards (NSPS), sources covered by national emission standards for hazardous air pollutants (NESHAP), sources regulated under the acid rain program (Title IV Acid Deposition Control) and the stratospheric ozone program (Title VI Stratospheric Ozone Protection), and sources covered by State implementation plans.

The permits must meet several requirements. They must contain all the applicable pollution control obligations of a source such as those imposed by State implementation plans and NSPS under the authority of Title I of the CAAA, under the air toxics program of Title III of the CAAA, under the acid rain program of Title IV of the CAAA, or under other applicable provisions of the CAAA. As necessary, the operating permits are to establish maximum source operating rates, maximum pollutant emission rates and/or pollutant control methods, and methods for determining compliance with the emission limits or control methods.

Emissions from many sources of particulate matter (PM) are regulated under the above titles of the CAAA including State implementation plans for particles with an aerodynamic diameter less than or equal to a nominal 10 microns (PM_{10}). Therefore, operating permits will be required for several categories of sources that emit PM. The processes and operations for lime manufacturing plants, which cause PM emissions, are discussed in this document and an operating permit for this source category is presented as a model that may be used in implementing a permit program. This model permit covers only the particulate matter aspects of this source category; it does not cover other pollutants for which the source may be regulated.

2.0 SOURCES OF EMISSIONS

Lime is the high-temperature product of the calcination of limestone. There are two kinds, high calcium lime (CaO) and dolomitic lime (CaO·MgO). Lime is manufactured in various types of kilns by one of the following reactions:

$$CaCO_3 + \text{heat} \cdot CO_2 + CaO \text{ (high calcium lime)} \quad (2\text{-}1)$$

$$CaCO_3 \cdot MgCO_3 + \text{heat} \cdot CO_2 + CaO \cdot MgO \text{ (dolomitic lime)} \quad (2\text{-}2)$$

In some lime plants, the resulting lime is reacted (slaked) with water to form hydrated lime.[1–4]

The basic processes in the production of lime are: (1) quarrying raw limestone; (2) preparing limestone for the kilns by crushing and sizing; (3) calcining limestone; (4) processing the lime further by hydrating; and (5) miscellaneous transfer, storage, and handling operations. A general material flow diagram for a lime manufacturing plant is given in Figure 2-1. Some of the operations shown may not be performed in all plants. Permit information on crushing and sizing is not covered in this document, but can be found under *Model Operating Permit to Control PM_{10} Emissions from Nonmetallic Mineral Processing Plants* (EPA-450/3-92-xxxb). Permit information on other non-process fugitive emission sources such as haul roads and storage piles can be found in *Model Operating Permit to Control PM_{10} Emissions from Non-Process Fugitive Emission Sources* (EPA-450/3-92-xxxl).[1–4]

The heart of a lime plant is the kiln. The most prevalent type of kiln in the United States is the rotary kiln, which is used in about 90 percent of all lime production. This kiln is a long, cylindrical, slightly inclined, refractory-lined furnace, through which the limestone and hot combustion gases pass countercurrently. Coal, oil, and natural gas may all be fired in rotary kilns. Product coolers and kiln feed preheaters of various types are commonly used to recover heat from the hot lime product and hot exhaust gases, respectively.[1–4]

The vertical, or shaft, kiln is the next most common type of kiln in the United States. This kiln is an upright heavy steel cylinder lined with refractory material. The limestone is charged at the top and is calcined as it descends

To be provided by the facility.

■ **FIGURE 2-1** **General process flow diagram for a lime manufacturing plant.**

slowly to discharge at the bottom of the kiln. A primary advantage of vertical kilns over rotary kilns is higher average fuel efficiency. The primary disadvantages of vertical kilns are their relatively low production rates and degradation to the quality of the lime product which occurs when coal is used as the fuel source for the process. Few vertical kilns have been installed in the United States in recent years because of high product quality requirements.[1–4]

Other much less common kiln types include rotary hearth and fluidized bed kilns. Both of these kiln types are capable of high production rates; neither require coal to operate. The "calcimatic" kiln, or rotary hearth kiln, is a circular shaped kiln with a slowly revolving, donut-shaped hearth. In fluidized bed kilns, finely divided limestone is brought into contact with hot combustion air in a turbulent zone, usually above a perforated grate. Because of the amount of lime carryover into the exhaust gases, dust collection equipment must be installed on fluidized bed kilns for process economy.[1–4]

About 10 percent of all lime produced is converted to hydrated or slaked lime. Atmospheric or pressure hydrators are used in this process. Atmospheric hydrators, the most commonly used type, are run in continuous mode to produce high calcium and normal dolomitic hydrates. Pressure hydrators, by contrast, produce only a completely hydrated dolomitic lime and operate only in batch mode. Generally, water sprays or wet scrubbers perform the hydrating process to prevent product loss. Following hydration, the product may be milled, after which it is conveyed to air separators for further drying and removal of coarse fractions.[1–4]

In the United States, lime plays a major role in chemical and metallurgical operations. Two of the largest uses are as steel flux and in alkali production. Lesser uses include construction, refractory, and agricultural applications.[1–4]

2.1 References

1. Lewis, C.J., and B.B. Crocker. The Lime Industry's Problem of Airborne Dust. *Journal of the Air Pollution Control Association*, 19(1):31–39, January 1969.
2. *Kirk–Othmer Encyclopedia of Chemical Technology*, 2nd edition, John Wiley & Sons, New York, 1967.
3. Vulcan-Cincinnati, Inc. *Screening Study for Emissions Characterization from Lime Manufacture.* Prepared for the U.S. Environmental Protection Agency under EPA Contract No. 68-02-0299, Cincinnati, OH, August 1974.
4. U.S. Environmental Protection Agency. *Standards Support and Environmental Impact Statement, Volume I: Proposed Standards of Performance for Lime Manufacturing Plants.* EPA-450/2-77-007a. Office of Air Quality Planning and Standards. Research Triangle Park, NC, April 1977.

3.0 MODEL PERMIT FOR LIME MANUFACTURING PLANTS

3.1 Introduction

Section 190 of Title I of the Clean Air Act Amendments of 1990 requires the Administrator of the U.S. Environmental Protection Agency to provide guidance to State and local agencies on categories of sources that contribute to nonattainment of the PM_{10} national ambient air quality standard. As part of that guidance, Section 3.2 provides specific conditions that may be included in operating permits that would apply to PM_{10} emission sources at lime manufacturing plants. These examples should be considered as guides for States to use in designing their own operating permit programs.

Individual facilities should be permitted on a case-by-case basis, and the emission limits included in the permit should be based on State implementation plan (SIP) regulations applicable in the area where the source is located. The mass and visible emission limits on an emission source and the operating conditions should be sufficiently detailed in the permit to allow a clear determination of whether the source is or is not in compliance. At a minimum, the permit should contain the following components:

- A fixed term, not to exceed five years [Section 502(b)(5)(B)];
- Limits and conditions to ensure compliance with all applicable requirements under the Act, including requirements of the applicable implementation plan [Section 504(a)];
- A schedule of compliance, which is defined as a schedule of remedial measures, including an enforceable sequence of actions or operations, leading to compliance with applicable requirements under the Act [Sections 504(a) and 501(3)];
- Inspection, entry, monitoring, compliance certification, recordkeeping, and reporting requirements to ensure compliance with the permit terms and conditions consistent with any monitoring regulations that EPA promulgates under Section 504(b) [Section 504(c)];
- A provision describing conditions under which any permit for a major source with a term of three or more years must be reopened to incorporate any new standard or regulation promulgated under the Clean Air Act [Section 502(b)(9)];

- Provisions under which the permit can be revised, terminated, modified, or reissued for cause [Section 502(b)(5)(D)]; and
- Provisions ensuring operational flexibility within a permit so that certain changes can be made without a permit revision, if no "modification" (as defined in Title I of the Act) would occur and the changes do not exceed the emissions allowable under the permit (whether expressed as a rate or in terms of total emissions) provided that a notice is sent to the permitting authority at least seven days in advance of the change [Section 502(b)(10)].

Examples of the following sections of a typical operation permit will be provided in a separate document:

- Source Identification
- Findings of Fact, Conclusions of Law, Decisions
- General Conditions
- Compliance Conditions
- Operational Flexibility

3.2 Specific Conditions*

The permittee shall limit particulate matter emissions as specified in Table 3-1.

■ **TABLE 3-1** Control Requirements for Lime Manufacturing Plant

Source	Emission Point No.[a]	Source Type[b]	Reference
Facility as a whole			Section 3.2.1
Rotary kiln	001	P	Section 3.2.2
Product cooler	002	P	Section 3.2.3
Hydrator	003	P	Section 3.2.4

[a]Emission points correspond to those shown in Figure 2-1.
[b]P = process emissions source.

3.2.1 Example of Specific Conditions for the Facility as a Whole

1. The permittee shall not permit or allow the stone feed rate to the kiln to exceed [____mg/hour (____tons/hour)] (maximum design capacity as specified in the permit application).
2. The permittee shall determine stone feed to the kiln by weight.

3.2.2 Example of Specific Conditions for Rotary Kiln (Emission Point 001)

1. The permittee shall not permit or allow to be discharged into the atmosphere any gases which contain particulate matter in excess of [____kg/mg (____lb/ton)] of stone feed or exhibit visible emissions greater than [____percent] opacity when exiting from a dry emission control device.
2. This section should contain operating parameters for the control equipment applied at this emission point (see Section 3.3 for example operating parameters).

**Note:* Specific control, performance, and reporting information contained in this model permit is provided solely for illustrative purposes and is not intended to establish industrial norms or national standards.

3.2.3 Example of Specific Conditions for Product Cooler (Emission Point 002)

1. The permittee shall not permit or allow to be discharged into the atmosphere any gases which contain particulate matter in excess of [____ g/dscm (____ gr/dscf)] or exhibit visible emissions greater than [____percent] opacity.
2. This section should contain operating parameters for the control equipment applied at this emission point (see Section 3.3 for example operating parameters).

3.2.4 Example of Specific Conditions for Lime Hydrator (Emission Point 003)

1. The permittee shall not permit or allow to be discharged into the atmosphere from any lime hydrator any gases which contain particulate matter in excess of [____g/dscm (____gr/dscf)] or exhibit visible emissions greater than [____percent] opacity.
2. This section should contain operating parameters for the control equipment applied at this emission point (see Section 3.3 for example operating parameters).

3.3 Control Equipment Operating Parameters

The parameters listed in this section are provided as examples of typical control equipment operating conditions that should be specified. There may be other important parameters not listed below.[1]

3.3.1 Fabric Filter Operating Parameters

Air/Cloth Ratio:________

Pressure Drop Across Baghouse:________in. H_2O (min)________in. H_2O (max)

Inlet Air Temperature:________(min)________(max)

Outlet Air Temperature:________(min)________(max)

Inlet Air Flow Rate:________

3.3.2 Electrostatic Precipitator Operating Parameters

Pressure Drop:________

Inlet Air Temperature:________(min)________(max)

Outlet Air Temperature:________(min)________(max)

Gas Velocity Through (ESP):________ft/sec

Conditioning Agent Additions:________gr/ft^3

Voltage and Current Readings for Each Transformer–Rectifier (T–R) Set:

Example: first 3 fields—3 at 50 KVA
(55 kV_p, 35 kV_{av}, 500 mA)
last 2 fields—4 at 64 KVA
(70 kV_p, 40 kV_{av}, 500 mA)

3.3.3 Mechanical Collectors Operating Parameters

Gas Flow Rate:________ft^3/min

Pressure Drop:________in. H_2O (min)________in. H_2O (max)

Inlet Velocity:________ft/sec

Inlet Air Temperature:________(min)________(max)

Outlet Air Temperature:________(min)________(max)

3.3.4 Scrubber Operating Parameters

Scrubber Liquor:________

Gas Flow Rate:________ft^3/min

Liquor Flow Rate:________gal/min

Gas Pressure Drop Across Unit:________in. H_2O (min)________in. H_2O (max)

Liquid Pressure Drop Across Unit:________in. psi. (min)________in. psi. (max)

Inlet Air Temperature:________(min)________(max)

Outlet Air Temperature:________(min)________(max)

Solids Content of Recirculated Scrubber Water:________gr/gal

3.4 Testing Requirements

Section 504(b) of the Clean Air Act, as amended, requires procedures and methods for determining compliance and for monitoring and analyzing pollutants regulated under the Act. An example of testing requirements that should be included in the permit is provided below.

1. The permittee shall perform a compliance test for all emission points from a stack or vent within 60 days of achieving the maximum production rate at the facility, but no later than 180 days after the initial startup, and annually thereafter. EPA Reference Method 5 or Method 5D shall be used to perform each compliance test on emission points from a stack. The sampling time and volume for each run shall be at least [____minutes] and [____dscm (____dscf)]. Testing of emissions shall be accomplished at a minimum of [____percent] of the permitted capacity. If testing is performed at a rate of less than [____percent] of the permitted capacity, operation is restricted to the process input rate of testing at such level, until a subsequent compliance test is performed at a minimum of [____percent] capacity.
2. EPA Reference Method 9 and the procedures in 40 CFR 60.11 shall be used to determine opacity.[2]
3. EPA Reference Method 22 shall be used to determine fugitive emissions from buildings.
4. At least [____calendar days] prior to performing a test, the owner or operator shall submit a test plan to the permitting authority that describes the test duration, test locations, test methods, source operation, and other parameters that may affect test results.

3.5 Monitoring Requirements

40 CFR 51.214(a) and Section 504(b) of the Clean Air Act, as amended, requires procedures and methods for monitoring to determine compliance with the State and Federal regulations as well as the emissions limits specified in the permit. Monitoring procedures and methods may include, but not be limited to, installation, maintenance, and operation of a continuous monitoring system, or daily visible emission observations, by a certified visible emissions observer, for at least three six-minute periods.

3.6 Recordkeeping Requirements

40 CFR 51.211 and Section 504(c) of the Clean Air Act, as amended, require permits to contain inspection, entry, monitoring, compliance certification, and reporting requirements to ensure compliance with permit terms and conditions. Provided below are examples of recordkeeping requirements.

1. Records shall be maintained for five years from the date of sample, measurement, report, or application.
2. The permittee shall obtain emission data for at least ____percent of the operational time of each monitored unit during each month of the calendar quarter. If the percentage of operational time of the opacity monitoring system is less than this amount, this may be considered the basis for declaring the installation in noncompliance with the applicable monitoring requirement. Furthermore, should the opacity monitoring system remain inoperative for more than ____days, the use of backup monitoring (e.g., monitoring process of control system parameters) will be required.[a]
3. Example of other potential recordkeeping requirements (to be defined by specific SIP requirement)[b]:
   a. The permittee shall maintain for five years any compliance test reports or quality assurance checks for the opacity monitoring system.

[a]The example cited in the text is the recordkeeping requirement as part of the PM-10 monitoring operations as required by a State implementation plan [40 CFR 51.214(c) and Appendix P].

[b]The State implementation plan requires the source to maintain records relating to the nature and amount of emissions, and other information as may be necessary to enable the permitting authority to determine whether the source is in compliance with applicable portions of the SIP [40 CFR 51.211].

3.7 Reporting Requirements

40 CFR 51.211 and Section 504(c) of the Clean Air Act, as amended, require the permits to contain reporting requirements to assure on a periodic basis compliance with permit terms and conditions, or with applicable portions of the control strategy as specified by the SIP. Provided below are examples of reporting requirements that should be consistent with the recordkeeping requirements.

1. The permittee shall submit quarterly excess emission reports within ____ days following the end of each calendar quarter. The reports will include, but not be limited to, the following information:[a]
   - **a.** The date and time of commencement and completion of each period of excess emissions.
   - **b.** The magnitude of the excess emissions.
   - **c.** The nature and cause of the excess emissions (if known) and the corrective action taken or preventative measures adopted.
   - **d.** The date and time of commencement and completion of any period which the emissions monitoring system was inoperative except for zero and span checks. The report shall identify the repairs or adjustments made to the system.
   - **e.** When no excess emissions have occurred during the quarter and the monitoring system had no period of downtime or did not require repairs or adjustments, an excess emissions report shall be filed stating such information.
2. The permittee shall report promptly [the State to define promptly] any deviations from permit requirements, including those attributable to upsets, and the report shall include the cause of such deviations and any corrective actions plan or preventive measures taken. Corrective actions may include a requirement for additional stack testing, or more frequent monitoring, or could trigger implementation of a corrective action plan.

[a]The examples cited in the text are the excess emission reporting requirements as specified by the Illinois General Air Rules [201.405].

3. The permittee shall report any deviations [the State to define specifically as possible those events that trigger additional reporting requirements] within ____hours of the event.
4. The permittee shall send ____copies of all reports required above to the [Permitting Authority Field Office].
5. Any document (including reports) required to be submitted by this permit shall be certified as being true, accurate, and complete by a responsible corporate official.

3.8 References

1. U.S. Environmental Protection Agency. *Technical Assistance Document: Recommended Recordkeeping Systems for Air Pollution Control Equipment—Part 1. Particulate Matter Controls.* EPA-340/1-86-021. Stationary Source Compliance Division. Washington, DC, August 1987.

2. *Standards of Performance for Lime Manufacturing Plants.* 40 CFR 60.344(b)(4).

INDEX